Graduate Texts in Physics

For further volumes:
http://www.springer.com/series/8431

Graduate Texts in Physics

Graduate Texts in Physics publishes core learning/teaching material for graduate- and advanced-level undergraduate courses on topics of current and emerging fields within physics, both pure and applied. These textbooks serve students at the MS- or PhD-level and their instructors as comprehensive sources of principles, definitions, derivations, experiments and applications (as relevant) for their mastery and teaching, respectively. International in scope and relevance, the textbooks correspond to course syllabi sufficiently to serve as required reading. Their didactic style, comprehensiveness and coverage of fundamental material also make them suitable as introductions or references for scientists entering, or requiring timely knowledge of, a research field.

Series Editors

Professor William T. Rhodes
Department of Computer and Electrical Engineering and Computer Science
    Imaging Science and Technology Center
Florida Atlantic University
777 Glades Road SE, Room 456
Boca Raton, FL 33431
USA
wrhodes@fau.edu

Professor H. Eugene Stanley
Center for Polymer Studies Department of Physics
Boston University
590 Commonwealth Avenue, Room 204B
Boston, MA 02215
USA
hes@bu.edu

Professor Richard Needs
Cavendish Laboratory
JJ Thomson Avenue
Cambridge CB3 0HE
UK
rn11@cam.ac.uk

János A. Bergou • Mark Hillery

Introduction to the Theory of Quantum Information Processing

 Springer

János A. Bergou
Department of Physics and Astronomy
Hunter College
City University of New York
New York, NY, USA

Mark Hillery
Department of Physics and Astronomy
Hunter College
City University of New York
New York, NY, USA

ISSN 1868-4513   ISSN 1868-4521 (electronic)
ISBN 978-1-4899-9843-9   ISBN 978-1-4614-7092-2 (eBook)
DOI 10.1007/978-1-4614-7092-2
Springer New York Heidelberg Dordrecht London

To Valéria, Attila, Miklós and Katalin
JB

To Carol
MH

Preface

These notes are the result of a one-semester graduate course that was first taught during the Spring 2003 Semester at the CUNY Graduate Center and has been offered several times since. The students in the courses were all physicists, so a familiarity with quantum mechanics at the first-year graduate level was assumed. The hope was that after taking the course, students could explore the original literature in the subject on their own.

The course covers a range of topics in quantum information but, given the limited amount of time, is not by any means exhaustive. We begin with the density matrix and its representations. Next we study entanglement, starting with Bell's inequalities and continuing with tests for entanglement, in particular, the Peres partial transposition test. It is also possible to quantify entanglement, and we show how this can be done for both pure and mixed states, finishing with a discussion of concurrence as a measure of entanglement for states of two qubits. Entanglement is a resource that can be used for quantum communication. Teleportation and dense coding are examples of this. Next, we consider quantum dynamics. In particular, we study generalized quantum dynamics that generalize the standard unitary evolution of quantum states. The Kraus representation of quantum maps is derived and applied to examples, such as the depolarizing channel. There are also certain kinds of maps that are impossible, such as the cloning map, a map that produces a perfect copy of an arbitrary input state.

We then move on to the study of quantum measurements. Just as quantum maps generalize the standard unitary evolution, positive operator valued measures (POVMs) generalize the standard projective measurements. Here we develop an extensive theory of generalized measurements that are described by POVMs. The problem of discriminating between two nonorthogonal quantum states provides a useful illustration of this type of measurement, and the two commonly employed strategies, the minimum-error strategy and the unambiguous state discrimination strategy, are discussed. These POVMs lead to a discussion of quantum cryptography. In particular, the B92 proposal and the original BB84 proposal are studied from this perspective. Many of the fascinating applications of quantum information theory in the area of quantum communication, such as secret sharing, rely on the impossibility of certain maps.

In quantum computation, the other major area of quantum information processing, consequences of the superposition principle are exploited. In the area of quantum algorithms, we focus primarily on the Deutsch–Jozsa algorithm, the Bernstein–Vazirani algorithm, the Grover search algorithm, and period finding. We also explore a technique that has been useful in finding new algorithms, the quantum walk. In a real quantum computation it is necessary to protect against errors, and for this quantum error-detecting codes are necessary. We develop the general theory of such codes and discuss some examples such as the Shor code and CSS codes.

We also have a chapter on quantum machines, devices that perform certain operations on quantum systems. These may be single purpose or programmable, and we discuss the limits on programmable machines. We conclude with an example of a programmable state discriminator, in which the states to be discriminated are provided as a program rather than hardwired into the machine.

This covers a lot of material, but it also leaves out a lot. In a single semester we cannot touch on subjects such as the applications of information theory to quantum information or the physical implementations of quantum information protocols, both of which are important subjects. We also do not treat the Shor algorithm for finding the prime factors of a number, not because it is not important but because it requires some background in number theory. When teaching a one-semester course, time constraints are a very real consideration, and we felt that an adequate presentation of the Shor algorithm and its background would take too much time. Our choice of subjects has been guided by the requirement of providing a firm foundation for further study and by our own interests as we have explored the field.

The chapters are completed with problems and a cursory list of the most relevant literature. The references are not meant to be exhaustive but to serve as a guide to further reading.

We should also mention two standard sources that we found useful in preparing the notes from which this book originated. One is *Quantum Computation and Quantum Information* by Michael Nielsen and Isaac Chuang. The second is the set of lecture notes by John Preskill for Physics 219 at Caltech, which can be found at http://www.theory.caltech.edu/people/preskill/ph229/. These cover some of the topics we discuss in more depth and also treat many topics that we do not. A more recent book, which can also supplement what we present here, is *Quantum Information* by Stephen Barnett.

Over the years, we benefitted from numerous discussions and close collaborations with many colleagues and friends. Among them we want to particularly thank Erika Andersson, Emilio Bagan, Stephen Barnett, Sam Braunstein, Vladimir Bužek, Luiz Davidovich, Berge Englert, Edgar Feldman, Ulrike Herzog, Igor Jex, Miguel Orszag, Daniel Reitzner, Wolfgang Schleich, Aephraim Steinberg, Mario Ziman, and M. Suhail Zubairy.

Finally, we are most grateful for the love and support of our families to whom this book is dedicated.

New York, NY, USA   János A. Bergou
New York, NY, USA   Mark Hillery

Contents

Chapter 1
Introduction

The field of quantum information encompasses the study of the representation, storing, processing, and accessing of information by quantum mechanical systems. The field grew from the investigations of the physical limits to computation initiated by Charles Bennett and Rolf Landauer. One of the first questions studied was whether quantum mechanics imposes any limits on what a computer can do, and it was shown by Richard Feynman that it does not. Earlier work by Paul Benioff had explored the possibilities of quantum Turing machines. Shortly after Feynman's work, David Deutsch realized that not only is quantum mechanics not a problem for computation, it can also be an advantage. The major breakthrough in the field was Peter Shor's factoring algorithm, which showed that a quantum computer can find the prime factors of integers in a time that scales as a polynomial of the size of the integer.

1.1 The Qubit

The basic unit of classical information is the bit, which can be 0 or 1. The corresponding object in quantum information is the qubit, which is a two-level quantum system. The two levels are often denoted by $|0\rangle$ and $|1\rangle$, which correspond to logical 0 and 1, respectively. Natural physical systems that can be used to represent a qubit are electronic or nuclear spins and the polarization of a photon. The key difference between a bit and a qubit is that the latter can exist in a superposition state

$$|\psi\rangle = \alpha|0\rangle + \beta|1\rangle, \tag{1.1}$$

while the former is definitely either 0 or 1. This leads to significant differences in what can be done with information represented by bits and that represented by qubits.

J.A. Bergou and M. Hillery, *Introduction to the Theory of Quantum Information Processing*, Graduate Texts in Physics, DOI 10.1007/978-1-4614-7092-2_1, © Springer Science+Business Media New York 2013

Fig. 1.1 Bloch sphere
representation of the qubit
in Eq. (1.2)

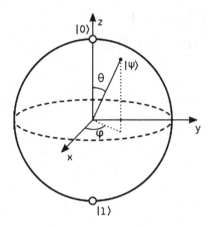

Fig. 1.1 Bloch sphere representation of the qubit in Eq. (1.2)

A convenient representation of the state of a qubit is given by the Bloch sphere. The state is parameterized by two angles, $0 \leq \theta \leq \pi$ and $0 \leq \phi < 2\pi$

$$|\psi\rangle = \cos(\theta/2)|0\rangle + e^{i\phi}\sin(\theta/2)|1\rangle. \tag{1.2}$$

It is represented by a point on the unit sphere whose polar angle is θ and whose azimuthal angle is ϕ. That is, the vector from the origin to the point representing the state makes an angle of θ with the z axis, and its component in the x-y plane makes an angle of ϕ with the x axis. The state $|0\rangle$ is the North Pole of the sphere, and the state $|1\rangle$ is the South Pole. The Bloch sphere of a qubit is shown in Fig. 1.1.

The Bloch sphere is often useful in illustrating how the state of a qubit is changed by a quantum map.

The state of n qubits is spanned by the tensor product basis

$$|0\rangle \otimes \ldots |0\rangle \otimes |0\rangle = |0\ldots00\rangle$$
$$|0\rangle \otimes \ldots |0\rangle \otimes |1\rangle = |0\ldots01\rangle$$
$$\vdots$$
$$|1\rangle \otimes \ldots |1\rangle \otimes |1\rangle = |1\ldots11\rangle. \tag{1.3}$$

Note that we have expressed these states as $|x\rangle$, where x is an n-digit binary number. The most general n-qubit state can be expressed as

$$|\Psi\rangle = \sum_{x=0}^{2^N-1} c_x |x\rangle. \tag{1.4}$$

1.2 Quantum Gates

Quantum gates are unitary operators that act on one or more qubits. They are unitary, because they represent the effect of some kind of time evolution on the state of the qubit, and the time development transformation is a unitary operator. Because of this fact, quantum gates must be reversible, that is, if we know the output state of the gate, we can infer what the input state was. This rules out quantum versions of certain classical gates. For example, the AND gate is a gate with a two-bit input and a one-bit output. The output is given by the product of the inputs, which implies that the output 0 can be produced by the inputs 00, 01, or 10. Thus, this gate is not reversible and, therefore, has no quantum version.

On the other hand, the NOT gate, which simply flips a bit, $0 \rightarrow 1$ and $1 \rightarrow 0$, is reversible, so a quantum version, which performs the operations, $|0\rangle \rightarrow |1\rangle$ and $|1\rangle \rightarrow |0\rangle$, exists. It has the following action on a general qubit state

$$\alpha|0\rangle + \beta|1\rangle \rightarrow \alpha|1\rangle + \beta|0\rangle. \tag{1.5}$$

If we represent the qubits state as a two-component column vector,

$$\alpha|0\rangle + \beta|1\rangle = \begin{pmatrix} \alpha \\ \beta \end{pmatrix}, \tag{1.6}$$

then the quantum NOT gate can be represented as the Pauli matrix, σ_x

$$\begin{pmatrix} 0 & 1 \\ 1 & 0 \end{pmatrix} \begin{pmatrix} \alpha \\ \beta \end{pmatrix} = \begin{pmatrix} \beta \\ \alpha \end{pmatrix}. \tag{1.7}$$

The X gate is represented in Fig. 1.2.

Here and in the following the left line represents the input qubit and the right line the output qubit.

There are also quantum gates that have no classical analogue. One particularly useful one is the Hadamard gate. It is a single-qubit gate and is represented by the circuit symbol in Fig. 1.3.

The Hadamard gate performs the following transformation:

$$H|0\rangle = \frac{1}{\sqrt{2}}(|0\rangle + |1\rangle)$$

$$H|1\rangle = \frac{1}{\sqrt{2}}(|0\rangle - |1\rangle). \tag{1.8}$$

Fig. 1.2 Circuit symbol for the NOT or X gate

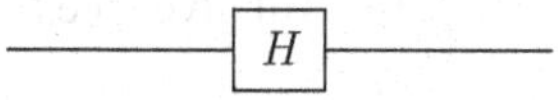

Fig. 1.3 Circuit symbol for the Hadamard gate

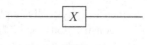

Fig. 1.4 Circuit symbol
for the C-NOT gate

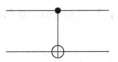

There is no classical analogue of this gate, because it takes the computational basis
states $\{|0\rangle, |1\rangle\}$ and maps them into superposition states. Note that $H^2 = I$, where I
is the identity operator.

A third important gate, which does have a classical analogue, is the Controlled-
NOT gate (or C-NOT gate for short), which is also known as the exclusive OR gate
(or XOR gate for short). It is a two-qubit gate and its circuit symbol is given in
Fig. 1.4.

The inputs are again on the left and the outputs on the right. The upper qubit is
called the control qubit and the lower one the target qubit. The state of the control
qubit is not changed by the gate, and the change in the state of the target qubit
depends on what the state of the control qubit is. In particular, if the control bit is
$|0\rangle$, nothing happens to the target qubit, but if the control bit is $|1\rangle$, then the target
qubit is flipped. In more detail, if the first qubit is the control and the second the
target, we have

$$|0\rangle|0\rangle \rightarrow |0\rangle|0\rangle \qquad |0\rangle|1\rangle \rightarrow |0\rangle|1\rangle$$
$$|1\rangle|0\rangle \rightarrow |1\rangle|1\rangle \qquad |1\rangle|1\rangle \rightarrow |1\rangle|0\rangle. \qquad (1.9)$$

From these relations the matrix elements of the transformation corresponding to the
C-NOT gate can be read out. This and the verification that the matrix is unitary are
left as a problem at the end of this chapter.

1.3   Quantum Circuits

At this stage, we are ready to introduce the circuit model of quantum computing.
In this model, qubits are represented by lines and quantum gates, i.e., unitary
operators, by their symbols. In particular, single-qubit gates are denoted by their
symbol on the line representing the qubit, two-qubit gates are denoted by their
symbol connecting two lines corresponding to two qubits, and so on. What makes
the circuit representation extremely useful is that the set of gates consisting of
the C-NOT and all single-qubit rotations is a universal set, which means that any
unitary transformation on any number of qubits can be constructed from them.
We shall not give the proof of this statement here. It is the reason that many
schemes for physically implementing quantum information protocols concentrate
on the construction of C-NOT gates. Instead of a more formal discussion, in the next
section, we will give an example illustrating how this type of description works.

1.4 The Deutsch Algorithm

The standard introductory example that is used to illustrate the fact that quantum information processing can be more powerful than classical information processing is Deutsch's problem. Consider a function, f, that maps the set $\{0,1\}$ to $\{0,1\}$. If $f(0) = f(1)$, then the function is constant, and if $f(0) \neq f(1)$, then we call f balanced. The problem is, given an unknown function, we want to determine whether it is constant or balanced. Classically we have to evaluate the function twice to determine this, but, using a quantum circuit, it is only necessary to evaluate it once. The quantum circuit that solves Deutsch's problem is shown in Fig. 1.5.

Let us see how this circuit works. The lines represent qubits, and the action proceeds from left to right. The gate labeled U_f is a two-qubit gate called an f-Controlled-NOT (f-CNOT). Like the C-NOT gate it has both a control (upper) qubit and a target (lower) qubit. The control qubit is not changed by the action of the gate, but the target bit has $f(x)$ added to it, modulo 2, where x is the value of the control bit. That is, if the input to the gate is $|x\rangle|y\rangle$, where x and y are the values of the control and target qubits, respectively, and are either 0 or 1, then the output is given by $|x\rangle|y + f(x)\rangle$. We have been given this gate, but we do not know what f is.

We shall now follow the qubits through the circuit. We start them in the state

$$|\Psi_0\rangle = |0\rangle_1 \frac{1}{\sqrt{2}}(|0\rangle_2 - |1\rangle_2), \tag{1.10}$$

where qubit 1 is the upper qubit and qubit 2 is the lower qubit. After the first gate the state is

$$|\Psi_1\rangle = \frac{1}{2}(|0\rangle_1 + |1\rangle_1)(|0\rangle_2 - |1\rangle_2), \tag{1.11}$$

and after the f-CNOT it is

$$|\Psi_2\rangle = \frac{1}{2}[|0\rangle_1(|0 + f(0)\rangle_2 - |1 + f(0)\rangle_2)$$
$$+ |1\rangle_1(|0 + f(1)\rangle_2 - |1 + f(1)\rangle_2). \tag{1.12}$$

Noting that

$$|0 + f(x)\rangle_2 - |1 + f(x)\rangle_2 = (-1)^{f(x)}(|0\rangle_2 - |1\rangle_2), \tag{1.13}$$

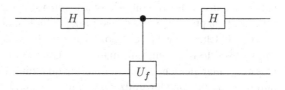

Fig. 1.5 Quantum circuit for Deutsch's problem

we have that

$$|\Psi_2\rangle = \frac{1}{2}[(-1)^{f(0)}|0\rangle_1 + (-1)^{f(1)}|1\rangle_1](|0\rangle_2 - |1\rangle_2). \qquad (1.14)$$

Finally, after passing through the second Hadamard gate, the state is

$$|\Psi_3\rangle = \frac{1}{2\sqrt{2}}\{|0\rangle_1[(-1)^{f(0)} + (-1)^{f(1)}]$$

$$+|1\rangle_1[(-1)^{f(0)} - (-1)^{f(1)}]\}(|0\rangle_2 - |1\rangle_2). \qquad (1.15)$$

Examining this expression, we see that if the function is constant, the first qubit is in the state $|0\rangle$ and if the function is balanced the first qubit is in the state $|1\rangle$. Therefore, by measuring the first qubit in the computational basis, we can determine whether f is constant or balanced. Note that the f-CNOT was used only once, so that the function was only evaluated once. The reason this procedure works is that what goes into the f-CNOT gate is a superposition of two input values, $|0\rangle$ and $|1\rangle$, and the function is evaluated on both of them at once (in the expression for $|\Psi_2\rangle$, both $f(0)$ and $f(1)$ appear). By carefully manipulating these values we can obtain information about the global properties of the function.

What we cannot do is obtain more than one value of the function if we evaluate it only once. Suppose we send the state $(1/\sqrt{2})(|0\rangle_1 + |1\rangle_1)|0\rangle_2$ into the f-CNOT gate. We have

$$\frac{1}{\sqrt{2}}(|0\rangle_1 + |1\rangle_1)|0\rangle_2 \rightarrow \frac{1}{\sqrt{2}}(|0\rangle_1|f(0)\rangle_2 + |1\rangle_1|f(1)\rangle_2). \qquad (1.16)$$

If we measure this state in the computational basis, we will obtain one of the values of f, and which one we obtain will be random. The measurement destroys the information about the other value of f that is present in the state. The lesson here is that we can use superpositions to evaluate a function on many different values of its argument simultaneously, but we have to be clever about how we use this information.

The example of Deutsch's problem tells us several things. The first is that there are gains to be had by representing information by quantum systems. The second is that finding how to produce these gains is far from straightforward. And last but not least, it shows that the final step in the algorithm is a measurement to read out the final state of the system. These are general features of all protocols and in the following chapters we will take an in-depth look at all of these ingredients.

First, in Chap. 2, we take a look on pure and mixed quantum states and how they can be used to represent information. In Chap. 3 we study their intrinsic quantum features, entanglement as a resource for quantum information and quantum computing, in particular. Generalized dynamics, operations and Kraus representation, is the subject of Chap. 4. The theory of quantum measurements, including generalized measurements, is presented in Chap. 5. Thus, the first five chapters contain what we can call the toolbox of quantum information theory. We then put these tools to good use in the rest of the book. In Chap. 6 we take a look at quantum communication; Chap. 7 deals with quantum computing (quantum algorithms, in particular). We included a rapidly evolving field, quantum machines, in Chap. 8. Finally, Chap. 9 deals with the main enemy: the inevitable influence of the environment, leading to decoherence and the protecting of quantum information (quantum error correction, in particular).

1.5 Problems

1. (a) Find the matrix corresponding to the Hadamard gate in the computational basis. (b) Find the matrix corresponding to the C-NOT gate in the two-qubit computational basis. (c) Check to see if the matrices in (a) and (b) are unitary.
2. Let us denote the unitary operator that implements a C-NOT gate by D_{ab}, where a is the control bit and b is the target bit. Let $|\psi\rangle = \alpha|0\rangle + \beta|1\rangle$ be a general qubit state, and let $|\pm x\rangle$ be the states given by $|\pm x\rangle = (|0\rangle \pm |1\rangle)/\sqrt{2}$.

   (a) We want to see what happens if the states $|\pm x\rangle$ are the input target bit states. Find the states $D_{ab}|\psi\rangle_a|\pm x\rangle_b$.
   (b) We can use the C-NOT to implement a one-parameter group of operations on a qubit probabilistically. Start by calculating

   $$D_{ab}|\psi\rangle_a(\cos\theta|+x\rangle_b + i\sin\theta|-x\rangle_b).$$

   Now measure the target qubit to see if it is in $|0\rangle_b$, or $|1\rangle_b$. Find the probability that it is in $|0\rangle_b$, and show that if it is, the control qubit is in the state $\exp(i\theta\sigma_z)|\psi\rangle_a$, where $\sigma_z|0\rangle = |0\rangle$ and $\sigma_z|1\rangle = -|1\rangle$.

3. The function that we considered in connection with Deutsch's algorithm is a special case of the so-called Boolean functions. A Boolean function $f^{(n)}$ maps the set of binary numbers $\{0, 1, \ldots, 2^n - 1\}$ to $\{0, 1\}$. We have shown that there are four different Boolean functions for $n = 1$, $f_1^{(1)}, f_2^{(1)}, f_3^{(1)}, f_4^{(1)}$. Two of them are constant and two of them are balanced.

   (a) Work out the truth table for each of these functions.
   (b) Find the corresponding f-C-NOT gate for each.
   (c) Show that they are unitary.

4. (a) A SWAP gate is a two-qubit gate that has the action $|\psi\rangle_a|\phi\rangle_b \to |\phi\rangle_a|\psi\rangle_b$. Show that a SWAP gate can be constructed from three C-NOT gates.
   (b) A Controlled-PHASE gate is a two-qubit gate, with qubit a as the control qubit and qubit b as the target qubit. The control qubit does not change, and if the control qubit is in the state $|0\rangle_a$ neither does the target qubit. If the control bit is in the state $|1\rangle_a$, then the gate acts as $|1\rangle_a|0\rangle_b \to |1\rangle_a|0\rangle_b$ and $|1\rangle_a|1\rangle_b \to -|1\rangle_a|1\rangle_b$. Show that a controlled-PHASE gate can be constructed from a C-NOT gate and two single-qubit gates.

References

1. D. Deutsch, Quantum theory, the Church-Turing principle, and the universal quantum computer. Proc. R. Soc. Lond. A **400**, 97 (1985)
2. D.P. DiVincenzo, Two qubit gates are universal for quantum computation. Phys. Rev. A **51**, 1015 (1995)
3. A. Barenco, C.H. Bennett, R. Cleve, D.P. DiVincenzo, N. Margolus, P. Shor, T. Sleator, J. Smolin, H. Weinfurter, Elementary gates for quantum computation. Phys. Rev. A **52**, 3457 (1995)

Chapter 2
The Density Matrix

We are going to require a more general description of a quantum state than that given by a state vector. The density matrix provides such a description. Its use is required when we are discussing an ensemble of pure states or when we are describing a subsystem of a larger system.

2.1 Ensembles and Subsystems

Let us look at ensembles first. Suppose that we have a collection of objects, some of which are in the quantum state $|\psi_1\rangle$, some of which are in $|\psi_2\rangle$, and so on. In particular, if we choose an object from the ensemble, the probability that it is in state $|\psi_j\rangle$ is p_j. We want to find the expectation value of some observable, Q, in this ensemble. We pick one of the objects in the ensemble and measure Q, pick another, and do the same. We repeat this process many times. If all of the objects were in the state $|\psi_j\rangle$, the expectation value of Q would be $\langle \psi_j|Q|\psi_j\rangle$, but in reality, objects with this state appear only with a probability p_j. Therefore, the expectation of Q in the ensemble is given by

$$\langle Q \rangle = \sum_j p_j \langle \psi_j|Q|\psi_j\rangle = \mathrm{Tr}(Q\rho), \tag{2.1}$$

where we have defined the operator ρ, which is the density matrix corresponding to the ensemble, to be

$$\rho = \sum_j p_j |\psi_j\rangle\langle \psi_j|. \tag{2.2}$$

Now let us look at subsystems. Suppose we have a large system composed of two subsystems, A and B. The Hilbert spaces for the quantum states of the subsystems are $\mathcal{H}_A$ and $\mathcal{H}_B$, so that the Hilbert space for the entire system is $\mathcal{H}_A \otimes \mathcal{H}_B$. Let $\{|m\rangle_A\}$ be an orthonormal basis for $\mathcal{H}_A$ and $\{|n\rangle_B\}$ be an orthonormal basis for $\mathcal{H}_B$.

J.A. Bergou and M. Hillery, *Introduction to the Theory of Quantum Information Processing*, Graduate Texts in Physics, DOI 10.1007/978-1-4614-7092-2_2, © Springer Science+Business Media New York 2013

Now if X_A is an observable on subsystem A, then the operator corresponding to it in the total Hilbert space is $X_A \otimes I_B$, where I_B is the identity on $\mathcal{H}_B$. If $|\Psi\rangle$ is the state of the entire system, then the expectation value of X_A is given by

$$
\begin{aligned}
\langle X_A \rangle &= \langle \Psi | X_A \otimes I_B | \Psi \rangle \\
&= \sum_m \sum_n \langle \Psi | X_A \otimes I_B (|m\rangle_A |n\rangle_B)(_A\langle m| \, _B\langle n|)|\Psi\rangle \\
&= \sum_m {}_A\langle m| \left(\sum_n {}_B\langle n|\Psi\rangle\langle\Psi|n\rangle_B \right) X_A |m\rangle_A .
\end{aligned}
\tag{2.3}
$$

If we now define

$$
\rho_A = \sum_n {}_B\langle n|\Psi\rangle\langle\Psi|n\rangle_B = \mathrm{Tr}_B(|\Psi\rangle\langle\Psi|),
\tag{2.4}
$$

then we have that

$$
\langle X_A \rangle = \mathrm{Tr}_A(\rho_A X_A).
\tag{2.5}
$$

The operator ρ_A is known as the reduced density operator for subsystem A, and it can be used to evaluate the expectation value of any observable that pertains only to subsystem A.

Now let us look at two examples. First, we have an ensemble in which half of the qubits are in the state $|0\rangle$ and the other half are in the state $|1\rangle$. The density matrix for this ensemble is

$$
\rho = \frac{1}{2}|0\rangle\langle 0| + \frac{1}{2}|1\rangle\langle 1|.
\tag{2.6}
$$

Suppose we want to find the expectation value of σ_z in this ensemble, where $\sigma_z|0\rangle = |0\rangle$ and $\sigma_z|1\rangle = -|1\rangle$. We have that

$$
\langle \sigma_z \rangle = \mathrm{Tr}(\sigma_z \rho) = 0.
\tag{2.7}
$$

Next, we have a two-qubit state

$$
|\Psi\rangle = \frac{1}{\sqrt{2}}(|0\rangle_A|1\rangle_B + |1\rangle_A|0\rangle_B),
\tag{2.8}
$$

and we would like to find the reduced density matrix for subsystem A. We have that

$$
\rho_A = \mathrm{Tr}(|\Psi\rangle\langle\Psi|) = \frac{1}{2}I_A.
\tag{2.9}
$$

Defining $\sigma_{zA} = \sigma_z \otimes I_B$, we have that

$$
\langle \sigma_{Az} \rangle = \mathrm{Tr}(\sigma_z \rho_A) = 0.
\tag{2.10}
$$

Note that if ρ is a one-dimensional projection, i.e., $\rho = |\psi\rangle\langle\psi|$, then for any observable Q,

$$\langle Q \rangle = \mathrm{Tr}(\rho Q) = \langle \psi | Q | \psi \rangle, \tag{2.11}$$

so that the density matrix corresponds to the system being in the state $|\psi\rangle$. If ρ is of this form we call it a pure state. If not, it is called a mixed state.

2.2 Properties

In order for an operator to be a density matrix, it must satisfy several properties. In fact, any operator satisfying these properties is a valid density matrix:

1. $\mathrm{Tr}(\rho) = 1$.
   This follows from the fact that

$$\mathrm{Tr}(\rho) = \mathrm{Tr}\left(\sum_j p_j |\psi_j\rangle\langle\psi_j|\right) = \sum_j p_j = 1. \tag{2.12}$$

2. A density matrix is hermitian, $\rho = \rho^\dagger$.
3. A density matrix is positive, $\langle\psi|\rho|\psi\rangle \geq 0$ for all $|\psi\rangle$.
   This follows from

$$\langle\psi|\rho|\psi\rangle = \sum_j p_j |\langle\psi|\psi_j\rangle|^2 \geq 0. \tag{2.13}$$

We also note that an operator is positive if and only if all of its eigenvalues are greater than or equal to zero, which implies that the eigenvalues of any density matrix must satisfy this property. In addition, because the trace of a density matrix is one and the trace is just the sum of the eigenvalues, we have that if λ_j is an eigenvalue of a density matrix, then $0 \leq \lambda_j \leq 1$.

We now want to use these requirements to find additional properties of the set of density matrices. The first is a simple way of identifying pure states.

Theorem. *The density matrix ρ is a pure state if and only if* $\mathrm{Tr}(\rho^2) = 1$.

Proof. If ρ is pure, then $\rho = |\psi\rangle\langle\psi|$, and $\rho^2 = \rho$. This immediately implies that $\mathrm{Tr}(\rho^2) = 1$. Now assume that $\mathrm{Tr}(\rho^2) = 1$. Because ρ is hermitian, we can express it as

$$\rho = \sum_j \lambda_j P_j, \tag{2.14}$$

where λ_j are the nonzero eigenvalues of ρ and P_j are the corresponding spectral projections. This immediately implies that

$$\rho^2 = \sum_j \lambda_j^2 P_j. \tag{2.15}$$

Denoting the rank of P_j by n_j we have that

$$\text{Tr}(\rho) = 1 \Rightarrow \sum_j \lambda_j n_j = 1$$

$$\text{Tr}(\rho^2) = 1 \Rightarrow \sum_j \lambda_j^2 n_j = 1, \tag{2.16}$$

and subtracting these two equations gives us

$$\sum_j (\lambda_j - \lambda_j^2) n_j = 0. \tag{2.17}$$

Because each eigenvalue is between 0 and 1, each term in the above sum is greater than or equal to zero, which further implies that each term must be equal to zero. The only way this can happen is if each λ_j is equal to zero or one, and we have assumed that $\lambda_j > 0$, so that $\lambda_j = 1$. The only way that this can be consistent with the fact that the sum of the eigenvalues times their multiplicities is one is if only one of them is nonzero, and this eigenvalue has a multiplicity of one. Therefore, ρ is equal to a rank-one projection, which means that it is a pure state. $\square$

2.3 Pure States and Mixed States of a Qubit

In the previous chapter we introduced the Bloch sphere as a convenient representation for state vectors of qubits, that is, qubit pure states. If we extend this representation to include the interior of the sphere, it can be used to represent mixed states of qubits as well. In order to see this we expand a general qubit density matrix, which is a 2×2 matrix, in terms of the identity matrix and the Pauli matrices, which form a complete basis for the space of a set of 2×2 matrices:

$$\rho = \frac{1}{2}(I + n_x \sigma_x + n_y \sigma_y + n_z \sigma_z). \tag{2.18}$$

This satisfies the condition $\text{Tr}(\rho) = 1$, and the fact that ρ is hermitian implies that n_x, n_y, and n_z are real. This equation implies that

$$\rho = \frac{1}{2} \begin{pmatrix} 1 + n_z & n_x - in_y \\ n_x + in_y & 1 - n_z \end{pmatrix}, \tag{2.19}$$

which further implies that $\det \rho = (1 - |\mathbf{n}|^2)/4$. The fact that ρ is positive means that its determinant must be greater than or equal to zero and, therefore, $1 \geq |\mathbf{n}|$. We represent the density matrix ρ by the vector $\mathbf{n}$, which lies in the unit ball.

We know that if ρ is a pure state, its corresponding vector will have its endpoint on the surface of the Bloch sphere. Let us show the converse. If $|\mathbf{n}| = 1$ then $\text{Tr}(\rho) = 1$

and $\det \rho = 0$. This implies that one of the eigenvalues of ρ is zero and the other is one. If $|u\rangle$ is the eigenvector with eigenvalue one, where $\|u\| = 1$, then $\rho = |u\rangle\langle u|$, and ρ is a pure state.

Given a qubit density matrix, ρ, we can easily find the vector corresponding to it. The identity $\mathrm{Tr}(\sigma_j \sigma_k) = \delta_{jk}$, where $j,k \in \{x,y,z\}$, gives us that

$$n_j = \mathrm{Tr}(\rho \sigma_j). \tag{2.20}$$

Most density matrices can correspond to many different ensembles. We give some examples. In the first example, we define the states

$$|\pm x\rangle = \frac{1}{\sqrt{2}}(|0\rangle \pm |1\rangle), \tag{2.21}$$

which are eigenstates of σ_x. Then we can write the density matrix of a maximally mixed state in two ways,

$$\rho = \frac{1}{2}I = \frac{1}{2}(|0\rangle\langle 0| + |1\rangle\langle 1|) = \frac{1}{2}(|+x\rangle\langle +x| + |-x\rangle\langle -x|). \tag{2.22}$$

The first decomposition corresponds to an ensemble in which half of the elements are in the state $|0\rangle$ and half in the state $|1\rangle$, and the second corresponds to an ensemble in which half of the elements are in the state $|+x\rangle$ and half in the state $|-x\rangle$. These ensembles are different, but they are described by the same density matrix.

In the second example, we define the states

$$|u_\pm\rangle = \frac{1}{\sqrt{4 \pm 2\sqrt{2}}}\left[(\sqrt{2} \pm 1)|0\rangle \pm |1\rangle\right]. \tag{2.23}$$

Then

$$\rho = \frac{1}{2}(|0\rangle\langle 0| + |+x\rangle\langle +x|) = \left(\frac{1}{2} + \frac{\sqrt{2}}{4}\right)\left(|u_+\rangle\langle u_+| + \left(\frac{1}{2} - \frac{\sqrt{2}}{4}\right)|u_-\rangle\langle u_-|\right), \tag{2.24}$$

again describing two different ensembles by the same density matrix.

In general, if ρ_1 and ρ_2 are density matrices, so is

$$\rho(\theta) = \theta\rho_1 + (1-\theta)\rho_2, \tag{2.25}$$

where $0 \leq \theta \leq 1$. This implies that the set of density matrices is convex. Most density matrices can be expressed as a sum of other density matrices in many different ways, and each of these decompositions will, in general, correspond to a different ensemble. The two examples above were just special cases of this general statement. This, however, is not true for pure sates; they have a unique decomposition. To see this suppose that $\rho = |\psi\rangle\langle\psi|$ is a pure state density matrix and that it can also be expressed as a convex sum of two other density matrices, $\rho(\theta) = \theta\rho_1 + (1-\theta)\rho_2$. Then if $|\psi_\perp\rangle$ satisfies $\langle\psi_\perp|\psi\rangle = 0$, then

$$0 = \langle\psi_\perp|\rho(\theta)|\psi_\perp\rangle = \theta\langle\psi_\perp|\rho_1|\psi_\perp\rangle + (1-\theta)\langle\psi_\perp|\rho_2|\psi_\perp\rangle. \tag{2.26}$$

Since both terms on the right-hand side are ≥ 0, it follows that

$$\langle\psi_\perp|\rho_1|\psi_\perp\rangle = \langle\psi_\perp|\rho_2|\psi_\perp\rangle = 0. \tag{2.27}$$

This equation is true for any vector orthogonal to $|\psi\rangle$. Therefore, $\rho_1 = \rho_2 = |\psi\rangle\langle\psi|$ and the representation of any pure state is unique. Pure states cannot be expressed as a sum of other density matrices. These are the only states with this property, because if ρ is mixed, it is given by $\rho = \sum_j p_j |\psi_j\rangle\langle\psi_j|$, which is just a convex sum of pure states.

2.4   Pure State Decompositions and the Ensemble Interpretation

Next we turn our attention to the ways in which a density matrix can be decomposed into pure states. The main result is summarized in the following:

Theorem. ρ *can be expressed as* $\sum_i p_i |\psi_i\rangle\langle\psi_i|$ *and* $\sum_i q_i |\phi_i\rangle\langle\phi_i|$ *iff*

$$\sqrt{p_i}|\psi_i\rangle = \sum_j U_{ij}\sqrt{q_j}|\phi_j\rangle,$$

where U_{ij} *is a unitary matrix and we "pad" whichever set of vectors is smaller with additional* 0 *vectors so that the sets have the same number of elements.*

Proof. Let $|\tilde{\psi}_i\rangle = \sqrt{p_i}|\psi_i\rangle$ and $|\tilde{\phi}_i\rangle = \sqrt{q_i}|\phi_i\rangle$. We first prove the if part, meaning that the condition is sufficient. To this end we suppose $|\tilde{\psi}_j\rangle = \sum_j U_{ij}|\tilde{\phi}_i\rangle$ for U_{ij} unitary. Then

$$\sum_i |\tilde{\psi}_i\rangle\langle\tilde{\psi}_i| = \sum_{i,j,k} U_{ij}U_{ik}^*|\tilde{\phi}_j\rangle\langle\tilde{\phi}_k| = \sum_{j,k}\left(\sum_i U_{ki}^\dagger U_{ij}\right)|\tilde{\phi}_j\rangle\langle\tilde{\phi}_k|$$

$$= \sum_{j,k}\delta_{jk}|\tilde{\phi}_j\rangle\langle\tilde{\phi}_k| = \sum_j |\tilde{\phi}_j\rangle\langle\tilde{\phi}_j|. \tag{2.28}$$

To prove the only if part, that the condition is also necessary, is considerably more work. We now suppose

$$\rho = \sum_i^{N_1} |\tilde{\psi}_i\rangle\langle\tilde{\psi}_i| = \sum_i^{N_2} |\tilde{\phi}_i\rangle\langle\tilde{\phi}_i|, \tag{2.29}$$

and assume $N_1 \geq N_2$. Since ρ is a positive operator, it has the spectral representation

$$\rho = \sum_{k=1}^{N_k} \lambda_k |k\rangle\langle k| = \sum_{k=1}^{N_k} |\tilde{k}\rangle\langle\tilde{k}|, \qquad (2.30)$$

where $\langle k|k'\rangle = \delta_{k,k'}$ and $|\tilde{k}\rangle = \sqrt{\lambda_k}|k\rangle$. First, we want to show that $|\tilde{\psi}_i\rangle$ lies in the subspace spanned by $\{|k\rangle\}$. To do this, let $\mathcal{H}_k$ denote the space spanned by $\{|k\rangle\}$. Suppose $|\psi\rangle \in \mathcal{H}_k^\perp$, then

$$\langle\psi|\rho|\psi\rangle = 0 = \sum_{i=1}^{N_1} |\langle\tilde{\psi}_i|\psi\rangle|^2. \qquad (2.31)$$

From here $\langle\tilde{\psi}_i|\psi\rangle = 0$ follows and so $|\tilde{\psi}_i\rangle \in (\mathcal{H}_k^\perp)^\perp = \mathcal{H}_k$. Therefore, we can express $|\tilde{\psi}_i\rangle$ as

$$|\tilde{\psi}_i\rangle = \sum_{k=1}^{N_k} c_{ik}|\tilde{k}\rangle. \qquad (2.32)$$

We can use this representation in Eq. (2.29) to obtain

$$\rho = \sum_{i=1}^{N_1} |\tilde{\psi}_i\rangle\langle\tilde{\psi}_i| = \sum_{k,k'=1}^{N_k} \left(\sum_{i=1}^{N_1} c_{ik}c_{ik'}^*\right)|\tilde{k}\rangle\langle\tilde{k}'| = \sum_{k=1}^{N_k} |\tilde{k}\rangle\langle\tilde{k}|. \qquad (2.33)$$

Since the operators $|\tilde{k}\rangle\langle\tilde{k}'|$ are linearly independent we have that $\sum_{i=1}^{N_1} c_{ik}c_{ik'}^* = \delta_{kk'}$. Thus c_{ik} form N_k orthogonal vectors of dimension N_1 and $N_1 \geq N_k$. In other words, c_{ik} ($k = 1,\ldots,N_k$) form the first N_k columns of a matrix containing N_1 rows that can be extended to an $N_1 \times N_1$ unitary matrix in the following way. Find $N_1 - N_k$ orthonormal vectors of dimension N_1 that are orthogonal to the vectors c_{ik} and call them c'_{ik}, where $i = 1,\ldots,N_1$ and $k = N_k+1,\ldots,N_1$. Obviously, the matrix

$$C_{ik} \equiv \begin{cases} c_{ik} & \text{for } k = 1,\ldots,N_k \\ c'_{ik} & \text{for } k = N_k+1,\ldots,N_1 \end{cases} \qquad (2.34)$$

for $i = 1,\ldots,N_1$ is an $N_1 \times N_1$ unitary matrix. If we introduce the vectors

$$\left(|\tilde{\psi}\rangle\right) = \begin{pmatrix} |\tilde{\psi}_1\rangle \\ \vdots \\ |\tilde{\psi}_{N_1}\rangle \end{pmatrix}, \qquad (2.35)$$

and

$$\left(|\tilde{k}_{N_1}\rangle\right) = \begin{pmatrix} |\tilde{k}_1\rangle \\ \vdots \\ |\tilde{k}_{N_k}\rangle \\ 0 \\ \vdots \\ 0 \end{pmatrix}, \qquad (2.36)$$

where the last $N_1 - N_k$ elements of $(|\tilde{k}_{N_1}\rangle)$ are 0, we can write

$$\left(|\tilde{\psi}\rangle\right) = \left(\begin{array}{ccc} \ddots & & \\ & C_{ik} & \\ & & \ddots \end{array}\right)\left(|\tilde{k}_{N_1}\rangle\right), \tag{2.37}$$

or, formally,

$$|\tilde{\psi}\rangle = C|\tilde{k}_{N_1}\rangle. \tag{2.38}$$

In an entirely similar way, we can show that $|\tilde{\phi}_i\rangle$ also lies in the subspace spanned by $\{|k\rangle\}$. Therefore, we can express $|\tilde{\phi}_i\rangle$ as

$$|\tilde{\phi}_i\rangle = \sum_{k=1}^{N_k} d_{ik}|\tilde{k}\rangle. \tag{2.39}$$

We can use this representation in Eq. (2.29) to obtain

$$\rho = \sum_{i=1}^{N_2} |\tilde{\phi}_i\rangle\langle\tilde{\phi}_i| = \sum_{k,k'=1}^{N_k}\left(\sum_{i=1}^{N_2} d_{ik}d_{ik'}^*\right)|\tilde{k}\rangle\langle\tilde{k}'| = \sum_{k=1}^{N_k}|\tilde{k}\rangle\langle\tilde{k}|. \tag{2.40}$$

Since the operators $|\tilde{k}\rangle\langle\tilde{k}'|$ are linearly independent we have that $\sum_{i=1}^{N_2} d_{ik}d_{ik'}^* = \delta_{kk'}$. Thus d_{ik} form N_k orthogonal vectors of dimension N_2 and $N_1 \geq N_2 \geq N_k$. In other words, d_{ik} ($k = 1,\ldots,N_k$) form the first N_k columns of a matrix containing N_2 rows that can be extended to an $N_2 \times N_2$ unitary matrix in the following way. Find $N_2 - N_k$ orthonormal vectors of dimension N_2 that are orthogonal to the vectors d_{ik} and call them d'_{ik}, where $i = 1,\ldots,N_2$ and $k = N_k + 1,\ldots,N_1$. Obviously, the matrix

$$D'_{ik} \equiv \begin{cases} d_{ik} & \text{for } k = 1,\ldots,N_k \\ d'_{ik} & \text{for } k = N_k + 1,\ldots,N_2 \end{cases} \tag{2.41}$$

for $i = 1,\ldots,N_2$ is an $N_2 \times N_2$ unitary matrix. Then we introduce the vector

$$\left(|\tilde{\phi}\rangle\right) = \left(\begin{array}{c} |\tilde{\phi}_1\rangle \\ \vdots \\ |\tilde{\phi}_{N_2}\rangle \\ 0 \\ \vdots \\ 0 \end{array}\right), \tag{2.42}$$

where the last $N_1 - N_2$ elements are 0 and unitarily extend D'_{ik} into an $N_1 \times N_1$ matrix D_{ik} by the following definition:

$$
D = \begin{pmatrix} \ddots & & \\ & D'_{ik} & 0 \\ & & \ddots \\ & 0 & I \end{pmatrix},
\tag{2.43}
$$

so that it is Eq. (2.41) for the first N_2 dimensions and identity for the remaining $N_1 - N_2$ dimensions. With these definitions we can now write

$$
\begin{pmatrix} \ddots \\ |\tilde{\phi}\rangle \end{pmatrix} = \begin{pmatrix} \ddots & & \\ & D'_{ik} & 0 \\ & & \ddots \\ & 0 & I \end{pmatrix} \begin{pmatrix} \ddots \\ |\tilde{k}_{N_1}\rangle \end{pmatrix},
\tag{2.44}
$$

or, formally,

$$
|\tilde{\phi}\rangle = D|\tilde{k}_{N_1}\rangle.
\tag{2.45}
$$

Comparing Eqs. (2.38) and (2.45), we finally obtain

$$
|\tilde{\psi}\rangle = CD^\dagger |\tilde{\phi}\rangle.
\tag{2.46}
$$

Since the matrix $U = CD^\dagger$ is unitary by construction, this completes the proof. $\square$

2.5 A Mathematical Aside: The Schmidt Decomposition of a Bipartite State

In the previous section we have looked at the possible decompositions of the density matrix in terms of convex sums of pure state density matrices. The decomposition is not unique, but the possible decompositions of the same density matrix are connected via the theorem proved in the previous section. Namely, the renormalized pure states, with appropriate zero vectors included if their numbers are different, are connected via a unitary transformation. Each of these decompositions gives rise to a different ensemble interpretation. The ensembles are not unique, but the various decompositions cannot be discriminated.

In this section we want to take a look at the other possible interpretation in which the mixed state density matrix emerges as the state of the subsystem of a larger system that itself is in a pure state. Therefore, we now examine the different ways in which a given density matrix ρ can be represented as the reduced density matrix for part of a pure bipartite state. To do this, we first need to derive the Schmidt decomposition of a bipartite state.

Let $|\psi\rangle_{AB} \in \mathcal{H}_A \otimes \mathcal{H}_B$, and $\{|u_i\rangle_A\}$ be an orthonormal basis for $\mathcal{H}_A$ and $\{|v_j\rangle_B\}$ be an orthonormal basis for $\mathcal{H}_B$. Then an arbitrary bipartite state can be expanded as a double sum over the product basis $\{|u_i\rangle|v_j\rangle\}$, as

$$|\psi\rangle_{AB} = \Sigma_{i,j} c_{ij} |u_i\rangle_A |v_j\rangle_B. \tag{2.47}$$

It is easy to see that this double sum expression can be written as a single sum,

$$|\psi\rangle_{AB} = \Sigma_i |u_i\rangle |\tilde{v}_i\rangle_B, \tag{2.48}$$

where we introduced $|\tilde{v}_i\rangle = \Sigma_j c_{ij} |v_j\rangle_B$. The price to pay is that $\{|\tilde{v}_i\rangle_B\}$ are not, in general, orthonormal. Therefore, it is somewhat surprising that for bipartite states, there exists a single sum expansion where only diagonal elements of a product basis, $\{|u_i\rangle|w_i\rangle\}$, enter.

In order to show this, suppose that $\{|u_i\rangle\}$ is the basis in which $\rho_A = \mathrm{Tr}_B(|\psi\rangle_{ABAB}\langle\psi|)$ is diagonal,

$$\rho_A = \Sigma_i \lambda_i |u_i\rangle\langle u_i|, \tag{2.49}$$

where $0 \le \lambda_i \le 1$. But we also have

$$\rho_A = \mathrm{Tr}_B[\Sigma_{(i,j)}(|u_i\rangle_A |\tilde{v}_i\rangle_B)(_A\langle u_j|_B\langle \tilde{v}_j|)] = \Sigma_{(i,j)} {}_B\langle \tilde{v}_j|\tilde{v}_i\rangle_B |u_i\rangle_{AA}\langle u_j|. \tag{2.50}$$

Therefore, we must have $_B\langle \tilde{v}_j|\tilde{v}_i\rangle_B = \delta_{ij}\lambda_i$ and hence $\{|\tilde{v}_i\rangle\}$ are orthogonal.

Let $\{|u_i\rangle_A \mid i = 1, \ldots, N, \text{where} N \le \dim\mathcal{H}_A\}$ correspond to nonzero values of λ_i and set $|w_i\rangle_B = \frac{1}{\sqrt{\lambda_i}}|\tilde{v}_i\rangle_B$. Hence $\{|w_i\rangle_B\}$ are orthonormal. Then

$$|\psi\rangle_{AB} = \Sigma_{i=1}^N \sqrt{\lambda_i} |u_i\rangle_A |w_i\rangle_B, \tag{2.51}$$

where $N \le \dim\mathcal{H}_A$ and by a similar argument $N \le \dim\mathcal{H}_B$. Note that

$$\rho_B = \mathrm{Tr}_A(|\psi\rangle_{ABAB}\langle\psi|) = \Sigma_{i=1}^N \lambda_i |w_i\rangle_{BB}\langle w_i|, \tag{2.52}$$

so that $\{|w_i\rangle\}$ are eigenstates of ρ_B having nonzero eigenvalues and ρ_A and ρ_B have the same nonzero eigenvalues. The double sum expansion in Eq. (2.47) always exists. It is somewhat surprising that the single sum expansion of Eq. (2.51), in terms of orthonormal basis vectors, also exists for bipartite systems. This later is called the Schmidt decomposition.

2.6 Purification, Reduced Density Matrices, and the Subsystem Interpretation

Equipped with the Schmidt decomposition, we now look at purifications. Suppose, we have a density matrix

$$\rho_A = \Sigma_{i=1}^N p_i |\psi_i\rangle_{AA} \langle \psi_i|, \tag{2.53}$$

where $|\psi_i\rangle \in \mathcal{H}_A$. We want to find a state $|\Phi\rangle_{AB} \in \mathcal{H}_A \otimes \mathcal{H}_B$ on a larger space so that

$$\rho_A = \mathrm{Tr}_B(|\Phi\rangle_{ABAB} \langle \Phi|). \tag{2.54}$$

$|\Phi\rangle_{AB}$ is called a purification of ρ_A.

One way to do this is to choose $\dim(\mathcal{H}_B) \geq N$ and let $\{|u_i\rangle\}$ be an orthonormal basis for $\mathcal{H}_B$. Then

$$|\Phi\rangle_{AB} = \Sigma_i \sqrt{p_i} |\psi_i\rangle_A |u_i\rangle_B \tag{2.55}$$

is called a purification of ρ_A.

Purifications are not unique. But if two purifications are in the same Hilbert space, we can still say something about their mutual relationship. Suppose we have two different states, $|\Phi_1\rangle_{AB}$ and $|\Phi_2\rangle_{AB}$, both of which are in $\mathcal{H}_A \otimes \mathcal{H}_B$ and both of which are purifications of ρ_A. How are they related? To answer this question, we use the Schmidt decomposition, $|\Phi_1\rangle_{AB} = \Sigma_k \sqrt{\lambda_k} |u_k\rangle_A |v_k\rangle_B$ and $|\Phi_2\rangle_{AB} = \Sigma_k \sqrt{\lambda_k} |u_k\rangle_A |w_k\rangle_B$. A part of both states, eigenvalues and eigenvectors of ρ_A, is the same. $\{|v_k\rangle_B\}$ and $\{|w_k\rangle_B\}$ form orthonormal sets, so there is at least one unitary operator on $\mathcal{H}_B$, which we call U_B, such that

$$|w_k\rangle_B = U_B |v_k\rangle_B. \tag{2.56}$$

Then $|\Phi_2\rangle_{AB} = (I_A \otimes U_B)|\Phi_1\rangle_{AB}$.

2.7 Problems

1. This problem combines elements from Chaps. 1 and 2 as it uses circuits with mixed states. We shall consider a complicated quantum circuit, one consisting of three qubits and four C-NOT gates. One of its uses is as a "quantum cloner." The operator for this circuit is given by $U = D_{ca}D_{ba}D_{ac}D_{ab}$ (remember that D_{ab} is a C-NOT gate with a as the control qubit and b as the target qubit).

   (a) Find $U(|\psi\rangle_a |\Psi_+\rangle_{bc}$, where

   $$|\Psi_+\rangle_{bc} = \frac{1}{\sqrt{2}}(|0\rangle_b |0\rangle_c + |1\rangle_b |1\rangle_c),$$

   and $U(|\psi\rangle_a |0\rangle_b| + x\rangle_c)$. What you should find is that in the first case $|\psi\rangle$ comes out of output a and in the second case it comes out of output b.

(b) Now find

$$|\Phi\rangle_{abc} = U|\psi\rangle_a(c_1|\Psi_+\rangle_{bc} + c_2|0\rangle_b|+x\rangle_c),$$

and find the condition on the constants c_1 and c_2 so that the input state is normalized. The idea here is that by combining the effects of the two input states in part (a), some of the information about $|\psi\rangle$ will end up in qubit a and some will end up in qubit b. How much ends up in each qubit depends on the values of c_1 and c_2.

(c) Find the reduced density matrixes for the outputs of qubits a and b, i.e., find

$$\rho_a = \mathrm{Tr}_{bc}(|\Phi\rangle_{abc\ abc}\langle\Phi|) \qquad \rho_b = \mathrm{Tr}_{ac}(|\Phi\rangle_{abc\ abc}\langle\Phi|)$$

In both cases your answer should be of the form

$$\rho = s|\psi\rangle\langle\psi| + \frac{1-s}{2}I,$$

where $0 \le s \le 1$. Find s in the case that $\rho_a = \rho_b$. Notice that what this device does is to produce two imperfect copies of the state $|\psi\rangle$.

2. The Schmidt representation for states of a bipartite system is extremely convenient, and so it is natural to ask if such a representation exists for tripartite systems. Unfortunately, the answer is no. Show that there exist three-qubit states that cannot be written in the form

$$|\Psi\rangle_{abc} = \sum_{j=0}^{1}\sqrt{\lambda_j}|u_j\rangle_a|v_j\rangle_b|w_j\rangle_c,$$

where $\{u_j|j=0,1\}$, $\{v_j|j=0,1\}$, and $\{w_j|j=0,1\}$ are orthonormal bases.

3. Suppose that Alice can prepare a density matrix only in the computational basis. She prepares a bipartite state of the form

$$\rho = \sum_{j,k=0}^{1} p_{jk}|j\rangle\langle j| \otimes |k\rangle\langle k|.$$

She sends one qubit to Bob and one qubit to Charlie. If Bob and Charlie do not measure in the computational basis, the correlations they can obtain are limited. Show that if they measure in the $|\pm x\rangle$ basis their results will be uncorrelated, that is, they are equally likely to get the same result as opposite results.

References

1. A. Peres, *Quantum Theory: Concepts and Methods* (Kluwer Academic Publishers, Dordrecht, 1995)
2. M. Nielsen, I. Chuang, *Quantum Computation and Quantum Information* (Cambridge University Press, Cambridge, 2010)

Chapter 3
Entanglement

3.1 Definition of Entanglement

We begin with some definitions. Consider a quantum state in a tensor product Hilbert space, $\mathcal{H} = \mathcal{H}_A \otimes \mathcal{H}_B$. A pure state is not entangled if it is of product form

$$|\psi\rangle_{AB} = |\phi_1\rangle_A \otimes |\phi_2\rangle_B, \tag{3.1}$$

otherwise, it is entangled. A density matrix, ρ_{AB}, is separable if it is a mixture of product states, i.e., if it is of the form

$$\rho_{AB} = \sum_i p_i \rho_{Ai} \otimes \rho_{Bi}, \tag{3.2}$$

where $0 \leq p_i \leq 1$, and $\sum_i p_i = 1$. If ρ_{AB} is not separable, it is entangled. For a pure state that is not entangled, measurements on systems A and B are not correlated. For a separable density matrix there are only classical correlations between measurements conducted on the two systems. As we shall see, entangled states can lead to much stronger correlations than are possible classically. Finally, we call a state maximally entangled if it is entangled and its reduced density matrices are proportional to the identity.

Let us briefly look at some two-qubit examples. The pure state $|0\rangle_A|0\rangle_B$ is not entangled, and neither is the density matrix

$$\rho_{AB} = \frac{1}{3}|0\rangle_{AA}\langle 0| \otimes |0\rangle_{BB}\langle 0| + \frac{2}{3}|1\rangle_{AA}\langle 1| \otimes |1\rangle_{BB}\langle 1|. \tag{3.3}$$

J.A. Bergou and M. Hillery, *Introduction to the Theory of Quantum Information Processing*, Graduate Texts in Physics, DOI 10.1007/978-1-4614-7092-2_3, © Springer Science+Business Media New York 2013

On the other hand, the so-called Bell states,

$$|\Phi_{\pm}\rangle_{AB} = \frac{1}{\sqrt{2}}(|01\rangle_{AB} \pm |10\rangle_{AB})$$

$$|\Psi_{\pm}\rangle_{AB} = \frac{1}{\sqrt{2}}(|00\rangle_{AB} \pm |11\rangle_{AB}), \tag{3.4}$$

are maximally entangled.

3.2 Bell Inequalities

Because entangled states can have correlations that go beyond what is possible classically, they are a valuable resource in quantum communication protocols, for example, as we shall see, in teleportation and dense coding. Before getting to these, however, let us see what is meant by nonclassical correlations, which means having a look at Bell inequalities.

These inequalities arose from a consideration of alternatives to quantum mechanics known as local hidden-variable theories. The idea behind them is that, unlike in quantum mechanics, observables have actual values but we do not know what they are, because they depend on some "hidden variables" about which we know nothing. In quantum mechanics, observables do not have values until we measure them. Bell inequalities show that under very general assumptions, hidden variables produce predictions that conflict with quantum mechanics. These can then be tested experimentally, and the experiments support quantum mechanics.

The basic setup for Bell inequalities consists of two observers, Alice and Bob, and a source that produces two-particle states. One particle is sent to Alice and the other to Bob. Alice can measure one of two observables for her particle, a_1 and a_2. These observables can each be either 1 or -1. Similarly, Bob can measure either b_1 or b_2, and these can also be either 1 or -1. The idea is to run this Gedankenexperiment many times and use the results to compute the quantities $\langle a_i b_j \rangle$.

Let us first see how a hidden-variable theory would describe this situation. The source produces, along with the particles, instruction sets that go with them. For example, one instruction set might say, if Alice measures a_1 she will get 1 if she measures a_2, she gets -1, and if Bob measures b_1 he gets -1 and if he measures b_2 he gets -1, or more briefly, $(a_1 = 1, a_2 = -1, b_1 = -1, b_2 = -1)$. We do not know which instruction set the source will produce, and so this, the instruction set, is our hidden variable. The adjective local is applied to this kind of a hidden-variable theory, because the instructions to Alice's particle do not depend on what Bob decides to measure. That is, the instruction set does not say something like, if Alice measures a_1, then she gets 1 if Bob measures b_1 and -1 if Bob measures b_2. We shall consider only local theories. We assume that each instruction set occurs with some probability. This is equivalent to assuming that we have a joint probability distribution for the variables, a_1, a_2, b_1, and b_2, which we shall denote as $P(a_1,a_2,b_1,b_2)$. We would then compute the expectation value $\langle a_1b_1 \rangle$, as

$$\langle a_1 b_1 \rangle = \sum_{a_1=-1}^{1} \cdots \sum_{b_2=-1}^{1} a_1 b_1 P(a_1,a_2,b_1,b_2). \tag{3.5}$$

We now want to consider the quantity

$$S = \langle a_1 b_1 \rangle + \langle a_1 b_2 \rangle + \langle a_2 b_1 \rangle - \langle a_2 b_2 \rangle$$

$$= \sum_{a_1=-1}^{1} \cdots \sum_{b_2=-1}^{1} [a_1(b_1+b_2) + a_2(b_1-b_2)] P(a_1,a_2,b_1,b_2). \tag{3.6}$$

Call the term in brackets multiplying the probability distribution X. We see that $X = a_1(b_1+b_2)$ if $b_1 = b_2$ and $X = a_2(b_1-b_2)$ if $b_1 = -b_2$. In both cases, $|X| = 2$, so that

$$|S| \leq 2 \sum_{a_1=-1}^{1} \cdots \sum_{b_2=-1}^{1} P(a_1,a_2,b_1,b_2) = 2. \tag{3.7}$$

This is a Bell inequality. Note that we can derive similar inequalities simply by interchanging a_1 and a_2, b_1 and b_2, or both.

Now let us describe the same experiment using quantum mechanics, and assume that we are measuring the spins of two spin-1/2 particles. Assume

$$a_1 = \sigma_{xa} \qquad a_2 = \sigma_{ya}$$

$$b_1 = \sigma_{xb} \qquad b_2 = \sigma_{yb}, \tag{3.8}$$

and that the source puts out particles in the state

$$|\Psi\rangle = \frac{1}{\sqrt{2}}(|00\rangle + e^{i\pi/4}|11\rangle), \tag{3.9}$$

where

$$\sigma_x|0\rangle = |1\rangle \qquad \sigma_y|0\rangle = i|1\rangle$$

$$\sigma_x|1\rangle = |0\rangle \qquad \sigma_y|1\rangle = -i|0\rangle. \tag{3.10}$$

Note that $|\Psi\rangle$ is an entangled state. We have that $\langle a_1b_1 \rangle$, $\langle a_1b_2 \rangle$, and $\langle a_2b_1 \rangle$ are all equal to $\sqrt{2}/2$ and $\langle a_2b_2 \rangle$ is equal to $-\sqrt{2}/2$. This gives us $S = 2\sqrt{2}$, which violates the Bell inequality.

From this we can conclude two things. First, quantum mechanics cannot be described by a local hidden-variable theory. Second, in the hidden-variable theory, the correlations came from a classical joint distribution function. Therefore, quantum mechanics can produce stronger correlations than classical systems can.

Next we want to study the connection of the Bell inequality to entanglement. We shall do this by showing that if $|\Psi\rangle$ is not entangled, then the Bell inequality will be satisfied. If $|\Psi\rangle$ is a product state then the expectation values appearing in the Bell inequality factorize, i.e., $\langle a_i b_j \rangle = \langle a_i \rangle \langle b_j \rangle$. Define $x_i = \langle a_i \rangle$ and $y_j = \langle b_j \rangle$, for $i, j = 1, 2$, where $-1 \le x_i \le 1$ and $-1 \le y_j \le 1$. Let us denote by R the region in the y_1, y_2 plane given by $\{-1 \le y_j \le 1 | j = 1, 2\}$. We then have that

$$S = x_1(y_1 + y_2) + x_2(y_1 - y_2). \tag{3.11}$$

Now suppose that $y_1 - y_2 = c > 0$, where $c \le 2$. This line intersects the boundary of R on the line $y_1 = 1$ at the point $y_1 = 1, y_2 = 1 - c$ and on the line $y_2 = -1$ at the point $y_1 = c - 1$, $y_2 = -1$. This implies that $c - 2 \le y_1 + y_2 \le 2 - c$. Similarly, if $y_1 - y_2 = c < 0$, where $c > -2$, then this line intersects the boundary of R on the line $y_1 = -1$ at the point $y_1 = -1, y_2 = -1 - c$ and on the line $y_2 = 1$ at the point $y_1 = c + 1$, $y_2 = 1$. This implies that $-c - 2 \le y_1 + y_2 \le 2 + c$. We can summarize both of these cases by the inequality, for $|c| \le 2$,

$$|c| - 2 \le y_1 + y_2 \le 2 - |c|. \tag{3.12}$$

We, therefore, have that if $y_1 - y_2 = c$, then $S = x_1(y_1 + y_2) + x_2 c$, and

$$-2 \le |x_1|(|c| - 2) - |x_2||c| \le S \le |x_1|(2 - |c|) + |x_2||c| \le 2. \tag{3.13}$$

Hence, we can conclude that for a pure state that is not entangled, the Bell inequality will be satisfied. This conclusion can be easily extended to separable states, because a separable state is just an incoherent superposition of product states, and for each of the product states the Bell inequality is satisfied.

Now, that we have seen what kind of states satisfy the Bell inequality, let us address the opposite end and find the maximum violation of this inequality that quantum mechanics can provide. This is given by Tsirelson's inequality. In order to derive the Tsirelson bound, we note the a_j and b_j are hermitian operators with eigenvalues ± 1 and hence $a_j^2 = b_j^2 = I$. Let us further define the operator $C = a_1 b_1 + a_1 b_2 + a_2 b_1 - a_2 b_2$. We then have

$$2\sqrt{2} - C = \frac{1}{\sqrt{2}}(a_1^2 + a_2^2 + b_1^2 + b_2^2) - C = \frac{1}{\sqrt{2}}\left(a_1 - \frac{b_1 + b_2}{\sqrt{2}}\right)^2$$

$$+ \frac{1}{\sqrt{2}}\left(a_2 - \frac{b_1 + b_2}{\sqrt{2}}\right)^2 \ge 0. \tag{3.14}$$

Therefore, $\langle C \rangle \le 2\sqrt{2}$. Similarly, by changing all the negative signs to positive ones in the above equation, one can show that $\langle C \rangle \ge -2\sqrt{2}$, and, since $S = \langle C \rangle$, we have that

$$|S| \le 2\sqrt{2}, \tag{3.15}$$

which is Tsirelson's inequality.

Table 3.1 Bob's possible operations and the resulting two-qubit state at Alice's site

Bob's operation	I	σ_x	σ_y	σ_z
Alice's state	$|\Phi_-\rangle$	$|\Psi_-\rangle$	$-i|\Psi_+\rangle$	$-|\Phi_+\rangle$

3.3 Representative Applications of Entanglement: Dense Coding and Teleportation

In this section we look at some interesting applications of entanglement that reveal its power as a resource for quantum information-related tasks.

3.3.1 Dense Coding

In this protocol two parties, traditionally called Alice and Bob, can communicate two bits of classical information by exchanging only one qubit. The key is entanglement, of course. We assume that Alice and Bob share an entangled pair of qubits in the state $|\Phi_-\rangle = \frac{1}{\sqrt{2}}(|01\rangle_{AB} - |10\rangle_{AB}$. Out of a pair of qubits in this state, Alice has one of the qubits, labeled A, in her possession while the other, labeled B, is in Bob's possession. Bob then performs one of four operations on his qubit and sends it back to Alice. The four operations and the resulting two-qubit state, now entirely at Alice's site, are listed in Table 3.1.

The point is that Alice now has one of four orthogonal states and she can distinguish them perfectly. After performing a measurement in the Bell basis, Alice will know with certainty which of the four operations Bob performed. Bob sent only one particle, a single qubit, to Alice, but Alice can perfectly distinguish among four classical alternatives, i.e., one (entangled) qubit carried two classical bits of information.

3.3.2 Teleportation

Alice has a qubit, say A_1 in some quantum state $|\psi\rangle$, in her possession. She wants to transfer the quantum state of her qubit A_1 onto Bob's qubit B. Alice may not even know what $|\psi\rangle$ is. Measuring $|\psi\rangle$ and transmitting the classical information that is the result will not work; it is not enough information to reconstruct the state.

In the teleportation procedure Alice and Bob share an entangled pair A_2, B in the state $|\psi\rangle_{A_2B} = \frac{1}{\sqrt{2}}(|01\rangle_{A_2B} - |10\rangle_{A_2B})$. The total state of the three qubits, the one whose state is to be teleported and the entangled pair, is then

Table 3.2 Alice's measurement outcomes and Bob's subsequent operations

Alice's measurement yields	$\lvert\Phi_+\rangle$	$\lvert\Phi_-\rangle$	$\lvert\Psi_+\rangle$	$\lvert\Psi_-\rangle$
Bob performs	σ_z	I (nothing)	$\sigma_z\sigma_x$	σ_x

$$
\begin{aligned}
\lvert\psi\rangle_{A_1}\lvert\psi\rangle_{A_2B} &= \frac{1}{\sqrt{2}}(\alpha\lvert0\rangle_{A_1}+\beta\lvert1\rangle_{A_1})(\lvert01\rangle_{A_2B}-\lvert10\rangle_{A_2B}) \\
&= \frac{1}{\sqrt{2}}(\alpha\lvert00\rangle_{A_1A_2}\lvert1\rangle_B - \alpha\lvert10\rangle_{A_1A_2}\lvert0\rangle_B \\
&\quad + \beta\lvert10\rangle_{A_1A_2}\lvert1\rangle_B - \beta\lvert11\rangle_{A_1A_2}\lvert0\rangle_B) \\
&= \frac{1}{2}\{\lvert\Phi_+\rangle_{A_1A_2}(-\sigma_z\lvert\psi\rangle_B) + \lvert\Phi_-\rangle_{A_1A_2}(-\lvert\psi\rangle_B) \\
&\quad + \lvert\Psi_+\rangle_{A_1A_2}(-\sigma_x\sigma_z\lvert\psi\rangle_B) + \lvert\Psi_-\rangle_{A_1A_2}(\sigma_x\lvert\psi\rangle_B)\}. \quad (3.16)
\end{aligned}
$$

The key is in the last line. When the total three-qubit state is decomposed in terms of the four Bell basis states of the two qubits of Alice, the state of Bob's qubit associated with each of these terms is related in a simple way to the state to be teleported. When Alice measures her state in the Bell basis, she tells Bob over a classical channel what she got, and then Bob can apply the appropriate operator to his qubit to recover Alice's state. The four possible outcomes of Alice's measurement and the operations Bob performs corresponding to each of the measurement results are listed in Table 3.2.

All information about $\lvert\psi\rangle$ is transferred to Bob; none is left with Alice. After teleportation Alice is left in possession of a Bell state. If someone else prepared the state of the original A_1 qubit for Alice, she will never learn its state in the process. Nevertheless, the state will be faithfully teleported to Bob's qubit B.

3.4 Conditions of Separability

How can we tell if a given density matrix is separable? Necessary and sufficient conditions are known to exist for the simplest particular cases only. In general, there are no known necessary and sufficient conditions to determine whether the state is separable or entangled. There are, however, some sufficient conditions.

One of them is Bell inequality. For two qubits, choose $a_1 = \vec{n}_1 \cdot \vec{\sigma}$, $a_2 = \vec{n}_2 \cdot \vec{\sigma}$, $b_1 = \vec{n}_3 \cdot \vec{\sigma}$, $b_2 = \vec{n}_4 \cdot \vec{\sigma}$, with $\vec{n}_j$ being unit vectors. If ρ_{AB} violates a Bell inequality for some choice of the unit vectors $\{\vec{n}_j | j = 1, \ldots, 4\}$, it is entangled. This is not a particularly strong criterion as there is a large class of entangled states that satisfies Bell inequalities.

A stronger and more general test was found by Peres, which is known as the positive partial transpose (PPT) criterion. Consider a density matrix on $\mathcal{H}_A \otimes \mathcal{H}_B$ of arbitrary dimensions. We have the density matrix elements in some product basis $\rho_{m\mu;n\nu} = {}_A\langle m| \otimes {}_B\langle \mu|\rho|n\rangle_A \otimes |\nu\rangle_B$.

The partial transposition of ρ is the density matrix with the matrix elements

$$\rho^{T_B}_{m\mu;n\nu} = \rho_{m\nu;n\mu}. \tag{3.17}$$

The operator ρ^{T_B} depends on the basis in which the transpose is defined, but its eigenvalues do not. We say a state is PPT if $\rho^{T_B} \geq 0$. A separable state is always PPT. This is because if ρ_{AB} is separable then $\rho^{T_B}_{AB} = \Sigma_i p_i \rho_{Ai} \otimes \rho^T_{Bi}$, and if $\rho_{Bi} \geq 0$, then $\rho^T_{Bi} \geq 0$.

Therefore, if a partial transpose is not positive, the state is entangled. Thus, the PPT condition is sufficient. For $2 \otimes 2$ (two-qubit) and $2 \otimes 3$ (qubit-qutrit) systems the converse is also true: if a state is entangled the partial transpose is not positive. Thus, for these systems, the PPT condition is also necessary.

As an example, consider the two-qubit state

$$\rho_{AB} = p|\Phi_-\rangle_{AB\,AB}\langle\Phi_-| + (1-p)|00\rangle_{AB\,AB}\langle 00|. \tag{3.18}$$

It can be shown that if $p \leq \frac{1}{\sqrt{2}}$ all Bell inequalities will be satisfied by this state.

Let us, however, apply the PPT condition to the same state. In the computational basis $\{|00\rangle, |01\rangle, |10\rangle, |11\rangle\}$, the above density matrix can be written as

$$\rho = \begin{pmatrix} 1-p & 0 & 0 & 0 \\ 0 & \frac{p}{2} & -\frac{p}{2} & 0 \\ 0 & -\frac{p}{2} & \frac{p}{2} & 0 \\ 0 & 0 & 0 & 0 \end{pmatrix}. \tag{3.19}$$

Its partial transpose with respect to B is

$$\rho = \begin{pmatrix} 1-p & 0 & 0 & -\frac{p}{2} \\ 0 & \frac{p}{2} & 0 & 0 \\ 0 & 0 & \frac{p}{2} & 0 \\ -\frac{p}{2} & 0 & 0 & 0 \end{pmatrix}. \tag{3.20}$$

The eigenvalues can be determined from the secular equation $\det(\rho^{T_B} - \lambda I) = 0$, which yields

$$\left(\frac{p}{2} - \lambda\right)^2 \left(\lambda^2 - (1-p)\lambda - \frac{p^2}{4}\right) = 0, \tag{3.21}$$

so that the eigenvalues are $\lambda_{1,2} = \frac{p}{2}$ and $\lambda_{3,4} = \frac{1}{2}[(1-p) \pm (1-2p+2p^2)^{1/2}]$. Three of them are obviously positive. The fourth one is $\lambda_4 = \frac{1}{2}\{(1-p) - [(1-p)^2 + p^2]^{1/2}\} < 0$ for $p > 0$. Therefore, for $p > 0$, the partial transpose is not positive and the state is entangled. Note that the Bell inequalities are not violated for $p \leq \frac{1}{\sqrt{2}}$, so the PPT condition is stronger than the condition of violating the Bell inequality.

Another way of detecting entangled states is by means of entanglement witnesses. An entanglement witness, W, is a hermitian operator that satisfies two properties. The first is that $\text{Tr}(\rho_s W) \geq 0$ for all separable density matrices, ρ_s. The second is that there is at least one entangled density matrix, ρ_e, such that $\text{Tr}(\rho_e W) < 0$. Since W is a hermitian operator, it is, at least in principle, an observable and can be measured. Entanglement witnesses provide a method of experimentally determining whether a state is entangled.

Constructing an entanglement witness for a state whose partial transpose is negative is straightforward. Suppose that ρ^{T_B} has a negative eigenvalue, λ_- with a corresponding eigenvector $|\eta\rangle$. Making use of the fact that for any two operators, X and Y on $\mathcal{H}_A \otimes \mathcal{H}_B$, $\text{Tr}(X^{T_B} Y) = \text{Tr}(X Y^{T_B})$, we have that

$$\text{Tr}\left(\rho(|\eta\rangle\langle\eta|)^{T_B}\right) = \text{Tr}\left(\rho^{T_B}(|\eta\rangle\langle\eta|)\right) = \lambda_- < 0. \tag{3.22}$$

On the other hand, for ρ_s separable,

$$\text{Tr}\left(\rho_s(|\eta\rangle\langle\eta|)^{T_B}\right) = \text{Tr}\left(\rho_s^{T_B}(|\eta\rangle\langle\eta|)\right) > 0, \tag{3.23}$$

because $\rho_s^{T_B}$ is a positive operator. Therefore, $(|\eta\rangle\langle\eta|)^{T_B}$ is an entanglement witness for the state ρ.

There are many other separability conditions that have been developed during the last few years, so our discussion of this subject will be far from complete. What we will do is cover a few conditions that can be used with continuous-variable systems. With these systems, because they are infinite dimensional, applying the partial transpose condition can be difficult. Hence, having simpler conditions can be useful. All of these conditions can be derived from the partial transpose condition, but our derivations will not explicitly make use of this condition.

Let us consider two particles on a line or, alternatively, two modes of the electromagnetic field. Each particle has a position operator, x_j, and a momentum operator, p_j, where $j = 1, 2$ and $[x_j, p_j] = i$. In the case of field modes, these would be the quadrature operators $x_j = (a_j^\dagger + a_j)/\sqrt{2}$ and $p_j = i(a_j^\dagger - a_j)/\sqrt{2}$, where a_j and $a_j^\dagger$, $j = 1, 2$, are the annihilation and creation operators for the modes. The commutation relations obeyed by x_j and p_j imply that $(\Delta x_j)(\Delta p_j) \geq 1/2$, where $(\Delta x_j)^2 = \langle x_j^2 \rangle - \langle x_j \rangle^2$, and similarly for $(\Delta p_j)^2$. Now define the two operators

$$u = |\alpha| x_1 + \frac{1}{\alpha} x_2$$

$$v = |\alpha| p_1 - \frac{1}{\alpha} p_2, \tag{3.24}$$

where α is a real number. What we will show is that for all separable states

$$(\Delta u)^2 + (\Delta v)^2 \geq \alpha^2 + \frac{1}{\alpha^2}. \tag{3.25}$$

That means that if this condition is violated for a particular state, that state is entangled. However, if the condition is satisfied, we can conclude nothing about the entanglement of the state. Violation of this inequality, then, is a sufficient condition for entanglement, but not a necessary one.

We now want to prove this statement. We assume that the density matrix is separable, so it can be expressed as

$$\rho = \sum_k p_k \rho_{1k} \otimes \rho_{2k}. \tag{3.26}$$

We than have that

$$(\Delta u)^2 + (\Delta v)^2 = \sum_k p_k(\langle u^2 \rangle_k + \langle v^2 \rangle_k) - \langle u \rangle^2 - \langle v \rangle^2$$

$$= \sum_k p_k \left(\alpha^2 \langle x_1^2 \rangle_k + \frac{1}{\alpha^2} \langle x_2^2 \rangle_k + \alpha^2 \langle p_1^2 \rangle_k + \frac{1}{\alpha^2} \langle p_2^2 \rangle_k \right)$$

$$+ 2\frac{\alpha}{|\alpha|} \sum_k p_k(\langle x_1 \rangle_k \langle x_2 \rangle_k - \langle p_1 \rangle_k \langle p_2 \rangle_k) - \langle u \rangle^2 - \langle v \rangle^2, \tag{3.27}$$

where expectation values with respect to $\rho_{1k} \otimes \rho_{2k}$ are denoted by a subscript k and expectation values with respect to the entire density matrix, ρ, do not have a subscript. Continuing

$$(\Delta u)^2 + (\Delta v)^2 = \sum_k p_k \left(\alpha^2 (\Delta x_1)_k^2 + \frac{1}{\alpha^2} (\Delta x_2)_k^2 + \Delta(p_1)_k^2 + \frac{1}{\alpha^2} (\Delta p_2)_k^2 \right)$$

$$+ \sum_k p_k \langle u \rangle_k^2 - \left(\sum_k p_k \langle u \rangle_k \right)^2$$

$$+ \sum_k p_k \langle v \rangle_k^2 - \left(\sum_k p_k \langle v \rangle_k \right)^2. \tag{3.28}$$

The Schwarz inequality implies that

$$\left(\sum_k p_k \langle u \rangle_k \right)^2 \leq \sum_k p_k \langle u \rangle_k^2, \tag{3.29}$$

and similarly for v. Therefore, we have that

$$(\Delta u)^2 + (\Delta v)^2 \geq \sum_k p_k \left(\alpha^2 (\Delta x_1)_k^2 + \frac{1}{\alpha^2} (\Delta x_2)_k^2 \right.$$

$$\left. + \Delta(p_1)_k^2 + \frac{1}{\alpha^2} (\Delta p_2)_k^2 \right). \tag{3.30}$$

Now the uncertainty relation between x_1 and p_1 implies that

$$(\Delta x_1)_k^2 + (\Delta p_1)_k^2 \geq (\Delta x_1)_k^2 + \frac{1}{4(\Delta x_1)_k^2} \geq 1, \tag{3.31}$$

and similarly for x_2 and p_2. Inserting these inequalities into Eq. (3.30) gives us the desired result.

The case $\alpha = 1$ gives a particularly simple result. In that case, we find that a state is entangled if

$$(\Delta(x_1 + x_2))^2 + (\Delta(p_1 - p_2))^2 < 2. \tag{3.32}$$

Noting that $[x_1 + x_2, p_1 - p_2] = 0$, both uncertainties can be made as small as we wish. What we see from the above inequality is that if they are sufficiently small, the state must be entangled.

Only a subset of entangled states will result in a violation of the inequality in Eq. (3.25). For example, the two-mode state $(|0\rangle_1|1\rangle_2 + |1\rangle_1|0\rangle_2)/\sqrt{2}$, that is, one photon in mode 1 and no photons in mode 2 plus no photons in mode 1 and one photon in mode 2, is an entangled state, but its entanglement will not be detected by Eq. (3.25). Consequently, there is room for more entanglement conditions. We will discuss one final condition, which will, in fact, show that the two-mode state we just mentioned is entangled.

We will prove this condition for an arbitrary system. Let A be an operator on $\mathcal{H}_A$ and B be an operator on $\mathcal{H}_B$. For a product state on $\mathcal{H}_A \otimes \mathcal{H}_B$, we have that

$$|\langle AB^\dagger \rangle| = |\langle A \rangle \langle B^\dagger \rangle| = |\langle AB \rangle| \leq \langle A^\dagger A B^\dagger B \rangle^{1/2}. \tag{3.33}$$

Now consider the density matrix for a general separable state given by $\rho = \sum_k p_k \rho_k$, where ρ_k is a density matrix corresponding to a pure product state and p_k is the probability of ρ_k. The probabilities satisfy the condition $\sum_k p_k = 1$. We then have that

$$|\langle AB^\dagger \rangle| \leq \sum_k p_k |\mathrm{Tr}(\rho_k AB^\dagger)|$$

$$\leq \sum_k p_k (\langle A^\dagger A B^\dagger B \rangle_k)^{1/2}, \tag{3.34}$$

where $\langle A^\dagger A B^\dagger B \rangle_k = \mathrm{Tr}(\rho_k A^\dagger A B^\dagger B)$. We can now apply the Schwarz inequality to obtain

$$|\langle AB^\dagger \rangle| \leq \left(\sum_k p_k \right)^{1/2} \left(\sum_k p_k \langle A^\dagger A B^\dagger B \rangle_k \right)^{1/2}$$

$$\leq (\langle A^\dagger A B^\dagger B \rangle)^{1/2}. \tag{3.35}$$

If a state violates this inequality, it is entangled. Note that this condition is very general, because we have not specified what A and B have to be. This condition can apply to finite dimensional spaces, infinite dimensional spaces, or a mixture of the two.

If we now go back to our two-mode state and choose $A = a_1$ and $B = a_2$, we find that for this state, $|\langle a_1 a_2^\dagger \rangle| = 1/\sqrt{2}$ and $\langle a_1^\dagger a_1 a_2^\dagger a_2 \rangle = 0$. This clearly violates the above inequality and thus proves that the state is entangled.

3.5 Entanglement Distillation and Formation

As we just saw in the examples of the previous sections, maximally entangled states of a pair of qubits are useful resources for several basic tasks in quantum communication, including dense coding and teleportation. In fact, they are so useful that they deserve their own name. If Alice and Bob share one maximally entangled two-qubit state, e.g., a singlet, then we say they share 1 ebit. Ebits, which are shared entanglement, are important resources and we now want to consider two other processes where they prove to be useful. These are:

- *Entanglement distillation.* Alice and Bob share n non-maximally entangled states. How many maximally entangled pairs (e.g., singlets or ebits) can they produce from them using only local operations and classical communication (LOCC)?
- *Entanglement formation.* Alice and Bob share n ebits and want to produce copies of some non-maximally entangled state $|\psi\rangle_{AB}$. How many copies of $|\psi\rangle_{AB}$ can they produce from them using only LOCC? Note that this is essentially the inverse of entanglement distillation, so we might as well just call it *entanglement dilution*.

3.5.1 Local Operations and Classical Communication [LOCC]

Of course, the first question we have to answer is: What is meant by LOCC? The meaning of classical communication is intuitively obvious and does not require further clarification. Local operations, on the other hand, need some explanation. They are operations performed by one party (either Alice or Bob) alone. The possibilities include:

 (i) Appending ancillary systems, not entangled with the other party
 (ii) Unitary operations
(iii) Orthogonal measurements
(iv) Throwing away part of the system

Note that the possibilities do not include the exchange of qubits.

Now, equipped with the concept of LOCC, we shall look at two simple examples that make use of these possibilities.

3.5.2 Entanglement Distillation: Procrustean Method

In this protocol, which is not optimal, Alice and Bob initially share the non-maximally entangled state $|\psi\rangle_{AB} = \cos\theta|00\rangle_{AB} + \sin\theta|11\rangle_{AB}$, where $\cos\theta > \sin\theta$, and want to extract the maximally entangled Bell state $|\psi_+\rangle = \frac{1}{\sqrt{2}}(|00\rangle_{AB} + |11\rangle_{AB})$. Note that $0 \le \theta \le \frac{\pi}{4}$. The protocol has three steps.

Step 1.   Alice appends ancilla qubit A' in state $|0\rangle_{A'}$, so that the total state becomes
$$|\psi\rangle_{AB} \otimes |0\rangle_{A'} = \cos\theta|00\rangle_{AA'} \otimes |0\rangle_B + \sin\theta|10\rangle_{AA'} \otimes |1\rangle_B.$$

Step 2.   Alice applies a unitary transformation U_A that performs the mapping

$$U_A|10\rangle_{AA'} = |10\rangle_{AA'},$$

$$U_A|00\rangle_{AA'} = \tan\theta|00\rangle_{AA'} + (1 - \tan^2\theta)^{1/2}|01\rangle_{AA'},$$

so that the total state becomes

$$(U_A \otimes I_B)(|\psi\rangle_{AB} \otimes |0\rangle_{A'}) = [\sin\theta|00\rangle_{AA'} + (1-2\sin^2\theta)^{1/2}|01\rangle_{AA'}] \otimes |0\rangle_B$$
$$+ \sin\theta|10\rangle_{AA'} \otimes |1\rangle_B$$
$$= \sqrt{2}\sin\theta|0\rangle_{A'} \otimes \frac{1}{\sqrt{2}}(|00\rangle_{AB} + |11\rangle_{AB})$$
$$+ (1 - 2\sin^2\theta)^{1/2}|1\rangle_{A'} \otimes |10\rangle_{AB}.$$

Step 3.   Alice measures the state of qubit A' and if she finds $|0\rangle_{A'}$, Alice and Bob keep the result because they just generated the ebit state $|\psi_+\rangle_{AB}$. Otherwise they throw away the result and repeat the steps.

The probability of success is $p_s = 2\sin^2\theta = 1 - \cos(2\theta)$. Thus, if Alice and Bob initially share n copies of $|\psi\rangle_{AB}$, the expected number of ebits resulting from this procedure is $n[1 - \cos(2\theta)]$.

3.5.3 Entanglement Formation

Again, the method is not optimal but demonstrates the power of entanglement. Initially, Alice and Bob share 1 ebit in the $|\phi_-\rangle_{AB}$ state and they want to generate the state $|\psi\rangle_{AB} = \cos\theta|00\rangle_{AB} + \sin\theta|11\rangle_{AB}$. The protocol has two steps:

Step 1.   Alice prepares the state $|\psi\rangle_{AA'} = \cos\theta|00\rangle_{AA'} + \sin\theta|11\rangle_{AA'}$ in her laboratory.

Step 2.   Alice uses the ebit (the singlet state shared with Bob) to teleport the state of particle A' to Bob.

The net result is that after the teleportation Alice and Bob share the state $|\psi\rangle_{AB} = \cos\theta|00\rangle_{AB} + \sin\theta|11\rangle_{AB}$.

3.6   Entanglement Measures

Up until now we spoke of entanglement in qualitative terms only. In this section we want to introduce entanglement measures that will tell us how entangled a state is. We will begin with pure states and gradually extend our measures to mixed states.

3.6.1   The von Neumann Entropy as an Entanglement Measure for Pure Bipartite States: A First Set of Properties

For a pure bipartite state $|\psi\rangle_{AB}$, we use the von Neumann entropy of one of the reduced density matrices as a measure of entanglement, E,

$$E(|\psi\rangle_{AB}) = S(\rho_A) = S(\rho_B). \tag{3.36}$$

Here $S(\rho) = -\text{Tr}(\rho\log_2\rho) = -\Sigma_i\lambda_i\log_2\lambda_i$ is the von Neumann entropy. Note that if $|\psi\rangle_{AB} = |\psi\rangle_A \otimes |\psi\rangle_B$ then $E(|\psi\rangle_{AB}) = 0$ since the von Neumann entropy of a pure state is 0.

The properties of E that make it a good entanglement measure can be listed as follows:

- **The entanglement of independent systems is additive.**

  Proof. If we have two independent pure state bipartite systems then tracing out one member of each system will leave the remaining two particles in independent mixed states, $\text{Tr}_{BB'}\{|\psi\rangle_{AB} \otimes |\psi'\rangle_{A'B'\,AB}\langle\psi| \otimes {}_{A'B'}\langle\psi'|\} = \rho_A \otimes \rho_{A'}$, where $\rho_A = \text{Tr}_B(|\psi\rangle_{ABAB}\langle\psi|)$ and $\rho_{A'} = \text{Tr}_{B'}(|\psi'\rangle_{A'B'A'B'}\langle\psi'|)$. So, now we need to show that $S(\rho_A \otimes \rho_{A'}) = S(\rho_A) + S(\rho_{A'})$. To this end we employ the diagonal representations, $\rho_A = \Sigma_n\lambda_n|n\rangle\langle n|$ and $\rho_{A'} = \Sigma_{n'}\lambda_{n'}|n'\rangle\langle n'|$, yielding $\rho_A \otimes \rho_{A'} = \Sigma_n\lambda_n\lambda_{n'}(|n\rangle\langle n|) \otimes (|n'\rangle\langle n'|)$. This finally gives $S(\rho_A \otimes \rho_{A'}) = -\text{Tr}\{\Sigma_{n,n'}\lambda_n\lambda_{n'}\log_2(\lambda_n\lambda_{n'})(|n\rangle\langle n|) \otimes (|n'\rangle\langle n'|)\} = -\Sigma_{n,n'}\lambda_n\lambda_{n'}(\log_2(\lambda_n) + \log_2(\lambda_{n'})) = S(\rho_A) + S(\rho_{A'})$. □

- **E is conserved under local unitary operations.**

  Proof. This follows from the cyclic property of the trace and can be shown in a straightforward manner. The most general local unitary operation can be written as $|\psi'\rangle_{AB} = U_A \otimes U_B |\psi\rangle_{AB}$ from which it immediately follows that $\rho' = U_A \rho_A U_A^{-1}$ and using the cyclic property of the trace operation we have that $S(\rho_A') = S(\rho_A)$. $\qquad\square$

- **E or, rather, the average value of E cannot be increased by LOCC.**
  The proof will be presented later (see Sects. 3.6.4–3.6.6).
- **Entanglement can be concentrated and distilled with asymptotic efficiency E using LOCC only** [C. Bennett, H. Bernstein, S. Popescu, and B. Schumacher, Phys. Rev. A **53**, 2046 (1996)]. Note that this is the best that can be done if E does not increase under LOCC.

  What this means is the following:

  (i) Alice and Bob share k copies of $|\psi\rangle_{AB}$ and from these produce n singlet pairs. Then, as $k \to \infty$, $\frac{n}{k} \to E(|\psi\rangle_{AB})$, using LOCC only.
  (ii) Alice and Bob share k copies of $|\Phi_-\rangle_{AB}$ (singlets) and from these produce n copies of $|\psi\rangle_{AB}$. Then, as $k \to \infty$, $\frac{k}{n} \to E(|\psi\rangle_{AB})$, using LOCC only.

3.6.2 A Useful Auxiliary Quantity: Relative Entropy and Klein's Inequality

To find further properties of the von Neumann entropy it is useful if we first introduce an auxiliary quantity, the so-called relative entropy. We define the relative quantum entropy of a state ρ relative to another state σ by

$$S(\rho \parallel \sigma) = \mathrm{Tr}(\rho \log \rho) - \mathrm{Tr}(\rho \log \sigma). \qquad (3.37)$$

An important feature of the relative entropy is that it is nonnegative, i.e., satisfies the Klein's inequality,

$$S(\rho \parallel \sigma) \geq 0. \qquad (3.38)$$

Proof. Let $\rho = \Sigma_i p_i |u_i\rangle\langle u_i|$ and $\sigma = \Sigma_i q_i |v_i\rangle\langle v_i|$ the diagonal representations of ρ and σ. Then $S(\rho \parallel \sigma) = \Sigma_i p_i (\log p_i - \langle u_i| \log \sigma |u_i\rangle)$ and since $\langle u_i| \log \sigma |u_i\rangle = \Sigma_j \log q_j \cdot |\langle u_i|v_j\rangle|^2$ we have that

$$S(\rho \parallel \sigma) = \Sigma_i p_i (\log p_i - \log q_j \cdot |\langle u_i|v_j\rangle|^2). \qquad (3.39)$$

Now, we want to use that $\log x$ is a concave function of x. This means that any straight line connecting two points $\log x_1$ and $\log x_2$ on the curve $\log x$ lies below $\log x$. Mathematically, the line $y = \log x_1 + \frac{\log x_2 - \log x_1}{x_2 - x_1}(x - x_1)$ for $x_1 \leq x \leq x_2$

lies below $\log x$. If we introduce $s = \frac{x-x_1}{x_1-x_2}$, this relationship can be written as $\log x_1 + s(\log x_2 - \log x_1) \leq \log[x_1 + s(x_2 - x_1)]$, or, after rearranging, we obtain $(1-s)\log x_1 + s\log x_2 \leq \log[(1-s)x_1 + sx_2]$.

If we introduce $r_i = \Sigma_i |\langle u_i | v_j \rangle|^2 q_j$ then the last inequality immediately gives $\Sigma_i |\langle u_i | v_j \rangle|^2 \cdot \log q_j \leq \log r_i$. Using this, in turn, in Eq. (3.39) gives

$$S(\rho \parallel \sigma) \geq \Sigma_i p_i \log \frac{p_i}{r_i}. \tag{3.40}$$

Since $\log x \ln 2 = \ln x \leq x - 1$, the right-hand side of this equation satisfies

$$\Sigma_i p_i \log \frac{p_i}{r_i} = -\Sigma_i p_i \log \frac{r_i}{p_i} \geq \Sigma_i p_i \left(1 - \frac{r_i}{p_i}\right) \cdot \frac{1}{\ln 2} = \Sigma_i (p_i - r_i) \cdot \frac{1}{\ln 2} = 0. \tag{3.41}$$

In light of Eq. (3.40), this just completes the proof of the Klein's inequality, Eq. (3.38). $\qquad\square$

3.6.3 The von Neumann Entropy: A Second Set of Properties

The Klein's inequality is a very powerful tool in studying the properties of the von Neumann entropy further. So, after this little mathematical detour, we return to the properties of $S(\rho)$:

- **For a d-dimensional system**

$$0 \leq S(\rho) \leq \log d. \tag{3.42}$$

Proof. The lower bound is obvious from the definition of the von Neumann entropy. We can obtain the upper bound by setting $\sigma = \frac{1}{d}I$ in the relative entropy. Then the Klein's inequality implies $S(\rho \parallel \sigma) = -S(\rho) - \Sigma_i \langle u_i | \rho \log(\frac{1}{d}I) | u_I \rangle = -S(\rho) + \log d$ and due to the Klein's inequality $-S(\rho) + \log d \geq 0$ which just gives us the upper bound for the von Neumann entropy. $\qquad\square$

- **Suppose p_i are probabilities, $|i\rangle$ orthonormal states for system A and $\{\rho_i\}$ a set of density matrices for system B. Then**

$$S(\Sigma_i p_i |i\rangle\langle i| \otimes \rho_i) = H(\{p_i\}) + \Sigma_i p_i S(\rho_i), \tag{3.43}$$

where $H(\{p_i\}) = -\Sigma_i p_i \log p_i$ is the Shannon entropy associated with the probability distribution $\{p_i\}$.

Proof. For fixed i, let $\{u_{ij}\}$ be the eigenstates of ρ_i with eigenvalues $\{\lambda_{ij}\}$. Using the basis $|i\rangle \otimes |u_{ij}\rangle$ to take the trace in the definition of the entropy gives

$$S(\Sigma_i p_i |i\rangle\langle i| \otimes \rho_i) = -\Sigma_i \Sigma_j p_i \lambda_{ij} \log(p_i \lambda_{ij})$$
$$= -\Sigma_i \Sigma_j p_i \lambda_{ij} (\log p_i + \log \lambda_{ij})$$
$$= -\Sigma_i p_i \log p_i + \Sigma_i p_i S(\rho_i), \qquad (3.44)$$

which just proves Eq. (3.43).   □

- **Subadditivity of the entropy.**
  Here we set out to prove the subadditivity property of the entropy, which states
  that if $\rho_A = \text{Tr}_B \rho_{AB}$ and $\rho_B = \text{Tr}_A \rho_{AB}$, then

$$S(\rho_{AB}) \leq S(\rho_A) + S(\rho_B). \qquad (3.45)$$

Proof. In order to prove this property we again apply Klein's inequality, this
time with $\rho = \rho_{AB}$ and $\sigma = \rho_A \otimes \rho_B$. We then have

$$S(\rho_{AB} \| \rho_A \otimes \rho_B) = \text{Tr}(\rho_{AB} \log \rho_{AB}) - \text{Tr}[\rho_{AB} \log(\rho_A \otimes \rho_B)] \geq 0, \qquad (3.46)$$

and also

$$\text{Tr}(\rho_{AB} \log \rho_A \otimes \rho_B) = \text{Tr}[\rho_{AB}(\log \rho_A \otimes I_B + I_A \otimes \log \rho_B)]$$
$$= \text{Tr}(\rho_A \log \rho_A) + \text{Tr}(\rho_B \log \rho_B). \qquad (3.47)$$

Putting now Eqs. (3.46) and (3.47) together gives

$$-S(\rho_{AB}) + S(\rho_A) + S(\rho_B) \geq 0, \qquad (3.48)$$

which proves the result.   □

A note is in order here. In a different context, the subadditivity inequality,
Eq. (3.45), is known as the triangle inequality. This is one of the most important
properties of quantities that are considered to be proper measures.

- Finally, we can put all of the previous properties together to prove a result
  that will actually say something about the effect of local measurement on
  entanglement.
  If $p_i \geq 0$ and $\Sigma_i p_i = 1$ and ρ_i are density operators, then

$$S(\Sigma_i p_i \rho_i) \geq \Sigma_i p_i S(\rho_i). \qquad (3.49)$$

Proof. We assume that ρ_i are density matrices of system A. Let us introduce an
auxiliary system B, with an orthonormal basis $\{|i\rangle\}$, and define $\rho_{AB} = \Sigma_i p_i \rho_i \otimes |i\rangle\langle i|$. From this definition it follows that $\rho_A = \Sigma_i p_i \rho_i$ and $\rho_B = \Sigma_i p_i |i\rangle\langle i|$ and also
$S(\rho_A) = S(\Sigma_i p_i \rho_i)$ and $S(\rho_B) = H(\{p_i\})$.

Applying the property given in Eq. (3.43) to this case we have that $S(\rho_{AB}) = H(\{p_i\}) + \Sigma_i p_i S(\rho_i)$. Applying the triangle inequality, Eq. (3.45), we finally have

$$H(\{p_i\}) + \Sigma_i p_i S(\rho_i) \leq S(\Sigma_i p_i \rho_i) + H(\{p_i\}), \qquad (3.50)$$

which proves the theorem.   □

We have seen so far that two of the four permissible local operations, namely, appending an additional system and applying local unitary transformations, have no effect on entanglement. Now, we are in the position to look at the effect of local measurements.

3.6.4 The Effect of Local Measurements on Entanglement

For the following considerations we again assume that Alice and Bob initially share the pure state $|\psi\rangle_{AB}$ and Alice performs a measurement on her particle. The possible outcomes of the measurement are labeled by k and the corresponding orthogonal projectors by P_k^A. In other words, P_k^A are the spectral projectors of the observable measured. She gets the result k with probability $p_k = {}_{AB}\langle\psi|P_k^A|\psi\rangle_{AB}$ and after this outcome was detected the state collapses to the unnormalized state $P_k^A|\psi\rangle_{AB\,AB}\langle\psi|P_k^A$.

If Alice does not communicate the result of her measurement to Bob, then Bob's density matrix cannot change, because otherwise superluminal communication would be possible. So, in this case, after the measurement, Bob's density matrix is

$$\rho_B = \mathrm{Tr}_A \left\{ \Sigma_k p_k P_k^A |\psi\rangle_{AB\,AB}\langle\psi|P_k^A \cdot \frac{1}{{}_{AB}\langle\psi|P_k^A|\psi\rangle_{AB}} \right\}$$

$$= \mathrm{Tr}_A \{ \Sigma_k P_k^A |\psi\rangle_{AB\,AB}\langle\psi|P_k^A \}. \qquad (3.51)$$

Clearly, $\mathrm{Tr}_A\{\Sigma_k P_k^A|\psi\rangle_{AB\,AB}\langle\psi|P_k^A\} = \mathrm{Tr}_A(|\psi\rangle_{AB\,AB}\langle\psi|) = \rho_B$, so we also see from the mathematics that Bob's density matrix is unchanged by the measurement.

If Alice does communicate her result to Bob, then Bob's density matrix can change, and so can the entanglement. In some cases it may even increase. The average entanglement will, however, always decrease. We define the average entanglement as $E = \Sigma_k p_k E(|\psi^{(k)}\rangle_{AB})$ where $|\psi^{(k)}\rangle_{AB} = ({}_{AB}\langle\psi|P_k^A|\psi\rangle_{AB})^{-1/2} P_k^A|\psi\rangle_{AB}$, since Alice and Bob share the state $|\psi^{(k)}\rangle_{AB}$ with probability p_k. Let $\rho_B^{(k)} = \mathrm{Tr}_A(|\psi^{(k)}\rangle_{AB\,AB}\langle\psi^{(k)}|)$, then $\Sigma_k p_k E(|\psi^{(k)}\rangle_{AB}) = \Sigma_k p_k S(\rho_B^{(k)})$. Now, using Eq. (3.51), we have that

$$\Sigma_k p_k \rho_B^{(k)} = \mathrm{Tr}_A \{ \Sigma_k p_k |\psi^{(k)}\rangle_{AB\,AB}\langle\psi^{(k)}| \} = \rho_B, \qquad (3.52)$$

and also $E(|\psi\rangle_{AB}) = S(\rho_B)$. Finally, using the subadditivity property, $S(\rho_B) \geq \Sigma_k p_k S(\rho_B^{(k)})$, we obtain

$$E(|\psi\rangle_{AB}) \geq \Sigma_k p_k E(|\psi^{(k)}\rangle). \tag{3.53}$$

Therefore, local measurements can only decrease the average entanglement shared by Alice and Bob.

3.6.5 Towards the Entanglement of Mixed States

Before looking at the last operation, throwing away part of the system, we have to define the entanglement of mixed states. This is because if we start with a pure state and throw away—that is to say trace out—part of it, we generally end up with a mixed state.

The basic idea of approaching this problem is the following. Let us start with a bipartite mixed state ρ_{AB} and decompose it into pure states, $\rho_{AB} = \Sigma_k p_k |\psi^{(k)}\rangle_{AB\,AB}\langle\psi^{(k)}|$. Alice and Bob want to create n copies of ρ_{AB}, how many singlet pairs will they need to do it? They can do it by creating np_k copies of $|\psi^{(k)}\rangle_{AB}$, for each k, and combining all the particles will erase the information of which singlet pair went with which value of k.

In order to create np_k copies of $|\psi^{(k)}\rangle_{AB}$, they will need $np_k E(|\psi^{(k)}\rangle_{AB})$ singlet pairs, so to create n copies of ρ_{AB}, they will need $\Sigma_k np_k E(|\psi^{(k)}\rangle_{AB})$ singlet pairs. We can then define the entanglement of ρ_{AB} as

$$E(\rho_{AB}) = \Sigma_k E(|\psi^{(k)}\rangle_{AB}). \tag{3.54}$$

However, there is a problem. The pure state decomposition of ρ_{AB} is not unique. We are interested in the minimum number of singlets to form ρ_{AB}, so we define the entanglement of formation of ρ_{AB} as

$$E(\rho_{AB}) = \inf \Sigma_k p_k E(\rho_{AB}), \tag{3.55}$$

where the infimum (largest lower bound) is taken over all possible pure state decompositions of ρ_{AB}. Finding this, except in some special cases, is hard.

3.6.6 The Effect of Throwing Away Part of the System Locally on Entanglement

Equipped with the concept of the entanglement of formation, which can be considered to be one possible entanglement measure for mixed states, we can finally turn our attention to addressing the problem: How throwing away part of a system locally affects entanglement?

In order to proceed with answering this last remaining question, let us assume that Alice and Bob initially share the pure state $|\psi\rangle_{AA'B}$. Alice now throws away system A', so that they are now sharing $\rho_{AB} = \text{Tr}_{A'}(|\psi\rangle_{AA'B} \cdot {}_{AA'B}\langle\psi|)$. Then the following theorem holds:

Theorem. *The entanglement of the composite system cannot be increased by throwing away part of the system locally,*

$$E(\rho_{AB}) \leq E(|\psi\rangle_{AA'B}). \tag{3.56}$$

Proof. We have $\rho_B = \text{Tr}_A(\rho_{AB}) = \text{Tr}_{AA'}(|\psi\rangle_{AA'B} \cdot {}_{AA'B}\langle\psi|)$ and also $E(|\psi\rangle_{AA'B}) = S(\rho_B)$. To calculate $E(\rho_{AB})$ we decompose it into pure states, $\rho_{AB} = \Sigma_k p_k |\psi_k\rangle\langle\psi_k|$. Then $E(\rho_{AB}) \leq \Sigma_k p_k E(|\psi_k\rangle\langle\psi_k|) = \Sigma_k p_k S(\rho_{Bk})$, where $\rho_B = \text{Tr}_A(\rho_{AB}) = \Sigma_k p_k \rho_{Bk}$. We also have that $\rho_B = \text{Tr}_A(\rho_{AB}) = \text{Tr}_k \rho_{Bk}$. Finally, from here, we have that

$$E(|\psi\rangle_{AA'B}) = S(\Sigma_k p_k \rho_{Bk}) \geq \Sigma_k p_k S(\rho_{Bk}) \geq E(\rho_{AB}), \tag{3.57}$$

which, when reading backwards, proves the theorem. □

What we have shown by all of the manipulations in this section is that if Alice and Bob start by sharing a pure state, they cannot increase their shared entanglement by LOCC. This result can be extended in a straightforward way to the case in which they are initially sharing a mixed state.

3.6.7 Bound Entanglement

It should be pointed out that not all entangled states can be distilled. That is, there are some entangled states from which a singlet state cannot be obtained by LOCC for any number of copies of the original state. Such states are called bound entangled. It can be shown that if a bipartite entangled state has a PPT, the state is bound entangled.

Let us give an example of a bound entangled state. In order to show that it is entangled, we will need an entanglement condition known as the range criterion. It states that if a density matrix ρ on $\mathcal{H}_A \otimes \mathcal{H}_B$ is separable, then there exists a family of product vectors, $|\psi_{Ak}\rangle \otimes |\psi_{Bk}\rangle$ such that it spans the range of ρ and the vectors $|\psi_{Ak}\rangle \otimes |\psi_{Bk}^*\rangle$ span the range of ρ^{T_B}, where $|\psi_{Bk}^*\rangle$ is the complex conjugate of $|\psi_{Bk}\rangle$, and the complex conjugation is performed in the same basis as the partial transpose.

In order to construct the bound entangled state we will need something called an unextendible product basis. This is a set of orthogonal product vectors in $\mathcal{H}_A \otimes \mathcal{H}_B$ that has fewer elements than the dimension of the space and is such that there is no product vector in the space that is orthogonal to all of the vectors in the set. An example for the case of two qutrits is

$$|v_0\rangle = \frac{1}{\sqrt{2}}|0\rangle(|0\rangle - |1\rangle) \quad |v_2\rangle = \frac{1}{\sqrt{2}}|2\rangle(|1\rangle - |2\rangle)$$

$$|v_1\rangle = \frac{1}{\sqrt{2}}(|0\rangle - |1\rangle)|2\rangle \quad |v_3\rangle = \frac{1}{\sqrt{2}}(|1\rangle - |2\rangle)|0\rangle$$

$$|v_4\rangle = \frac{1}{3}(|0\rangle + |1\rangle + |2\rangle)(|0\rangle + |1\rangle + |2\rangle). \tag{3.58}$$

Now define the projection $P = \sum_{j=0}^{4}|v_j\rangle\langle v_j|$. The claim is that the density matrix

$$\rho = \frac{1}{4}(I - P), \tag{3.59}$$

is a bound entangled state. First, it is entangled due to the range criterion. If there were a product vector in the range of ρ that would mean that the unextendible product basis could be extended, which it cannot. Hence, by the range criterion, ρ cannot be separable, so it must be entangled. The next step is to show that ρ^{T_B} is positive. In this case, $\rho = \rho^{T_B}$, so that ρ^{T_B} is clearly positive. Therefore, by the result mentioned in the first paragraph of this section, ρ is bound entangled.

3.7   Concurrence

In the case of two qubits it is possible to find the entanglement of formation of a general state explicitly. In order to do this, we have to introduce a quantity called concurrence. However, before we can define the concurrence, we must first define the so-called tilde state for a bipartite qubit pure state $|\psi\rangle_{AB}$, as

$$|\tilde{\psi}\rangle_{AB} = (\sigma_y \otimes \sigma_y)|\psi^*\rangle. \tag{3.60}$$

In this expression $|\psi^*\rangle$ is the complex conjugate of $|\psi\rangle$ in the standard basis, i.e., if $|\psi\rangle = \sum_{j,k=0}^{1} c_{jk}|j\rangle|k\rangle$, then $|\psi^*\rangle = \sum_{j,k=0}^{1} c_{jk}^*|j\rangle|k\rangle$.

The concurrence $C(|\psi\rangle_{AB})$ of $|\psi\rangle_{AB}$ is then defined as

$$C(|\psi\rangle) = |\langle\psi|\tilde{\psi}\rangle|. \tag{3.61}$$

Why does this quantity have anything to do with entanglement? Let us look at the single-qubit state, $|\psi\rangle = \alpha|0\rangle + \beta|1\rangle$, and its conjugate, $|\psi^*\rangle = \alpha^*|0\rangle + \beta^*|1\rangle$. Using $\sigma_y|0\rangle = i|1\rangle$ and $\sigma_y|1\rangle = -i|0\rangle$, we get $|\tilde{\psi}\rangle = \sigma_y|\psi^*\rangle = i(\alpha^*|1\rangle - \beta^*|0\rangle)$ from which

$$\langle \psi | \tilde{\psi} \rangle = 0 \qquad (3.62)$$

follows directly for single-qubit states. Let us now consider the Schmidt decompositions of $|\psi\rangle_{AB}$ and $|\tilde{\psi}\rangle_{AB}$, $|\psi\rangle_{AB} = \Sigma_{j=1}^{2} \sqrt{\lambda_j} |u_j\rangle_A |v_j\rangle_B$ and $|\tilde{\psi}\rangle_{AB} = \Sigma_{j=1}^{2} \sqrt{\lambda_j} |\tilde{u}_j\rangle_A |\tilde{v}_j\rangle_B$. In view of the orthogonality property of single-qubit states, Eq. (3.62), we have that $|\tilde{u}_1\rangle \propto |u_2\rangle$ and $|\tilde{u}_2\rangle \propto |u_1\rangle$, so $\langle \psi | \tilde{\psi} \rangle = \sqrt{\lambda_1 \lambda_2} (\langle u_1 | \tilde{u}_2 \rangle \langle v_1 | \tilde{v}_2 \rangle + \langle u_2 | \tilde{u}_1 \rangle \langle v_2 | \tilde{v}_1 \rangle)$. Setting $|u_1\rangle = \alpha |0\rangle + \beta |1\rangle$ and $|u_2\rangle = e^{i\phi_A} (\beta^* |0\rangle - \alpha^* |1\rangle)$ gives $\langle u_1 | \tilde{u}_2 \rangle = ie^{-i\phi_A}$ and $\langle u_2 | \tilde{u}_1 \rangle = -ie^{-i\phi_A}$ and, similarly, $\langle v_1 | \tilde{v}_2 \rangle = ie^{-i\phi_B}$ and $\langle v_2 | \tilde{v}_1 \rangle = -ie^{-i\phi_B}$, so all of the relevant inner products are simple phase factors. This leads to $\langle \psi | \tilde{\psi} \rangle = \sqrt{\lambda_1 \lambda_2} (-2e^{-i(\phi_A + \phi_B)})$, which finally yields

$$C(|\psi\rangle) = 2\sqrt{\lambda_1 \lambda_2} \qquad (3.63)$$

for the concurrence, with $\lambda_1 + \lambda_2 = 1$. The state is maximally entangled when $\lambda_1 = \lambda_2 = \frac{1}{2}$, giving the maximum value of $C = 1$. For a product state, on the other hand, $\lambda_1 = 0$ or $\lambda_2 = 0$, giving $C = 0$, so C is a monotonically increasing function of entanglement.

The entanglement of $|\psi\rangle_{AB}$ can be expressed as a function of C,

$$E(|\psi\rangle_{AB}) = \mathcal{E}(C(|\psi\rangle_{AB})), \qquad (3.64)$$

where

$$\mathcal{E}(C) = h\left(\frac{1 + \sqrt{1 - C^2}}{2} \right), \qquad (3.65)$$

and

$$h(x) = -x\log x - (1 - x)\log(1 - x) \qquad (3.66)$$

is the binary entropy associated with the probability distribution $\{x, 1 - x\}$.

Finally, let us look at mixed states. Suppose $\rho = \Sigma_k p_k |\psi_k\rangle\langle\psi_k|$, then

$$C(\rho) = \inf \Sigma_k p_k C(|\psi_k\rangle), \qquad (3.67)$$

where the infimum is taken over all possible pure state decompositions of ρ. The function $\mathcal{E}(C)$ is monotonically increasing, so

$$\mathcal{E}(C(\rho)) = \inf \mathcal{E}(\Sigma_k p_k C(|\psi_k\rangle)), \qquad (3.68)$$

and it is also convex; therefore,

$$\inf \mathcal{E}(\Sigma_k p_k C(|\psi_k\rangle)) \leq \inf \Sigma_k p_k \mathcal{E}(C(|\psi_k\rangle)) = E(\rho). \qquad (3.69)$$

From here it follows that

$$\mathcal{E}(C(\rho)) \leq E(\rho). \qquad (3.70)$$

To close this long chapter we list two more properties of the concurrence without presenting their proofs [W. Wooters, PRL 80, 2245 (1998)]:

1. This last inequality is actually an equality.
2. There is an explicit formula for $C(\rho)$. Let us first define ρ^* as the complex conjugate of ρ in the standard basis and then $\tilde{\rho} = (\sigma_y \otimes \sigma_y)\rho^*(\sigma_y \otimes \sigma_y)$. Let λ_i for $i = 1,\ldots,4$ denote the square roots of the eigenvalues of $\rho\tilde{\rho}$, arranged in decreasing order. Then

$$C(\rho) = \max\{0, \lambda_1 - \lambda_2 - \lambda_3 - \lambda_4\}. \tag{3.71}$$

3.8   Problems

1. Wigner's inequality is another inequality that will be satisfied by a local hidden-variable theory but can be violated by quantum mechanics. In order to derive it consider the following situation. A source sends one particle to Alice and one to Bob. Alice measures one of three observables, a_j, $j = 1,2,3$, and Bob also measures one of three, b_j, $j = 1,2,3$. Each of these observables gives the value 1 or -1. The source has the property that whenever Alice and Bob measure corresponding observables, that is, the value of j is the same, then the results are anticorrelated. For example, if Alice measures a_1 and Bob measures b_1, then they will never get the same result, but if Alice measures a_1 and Bob measures b_2, then they can get any result. Let $p(a_j = m, b_k = n)$ denote the probability that when Alice measures a_j and Bob measures b_k, Alice gets the value m and Bob gets n. Wigner's inequality states that if this source is described by a local hidden-variable theory, then

$$p(a_1 = 1, b_2 = 1) + p(a_2 = 1, b_3 = 1) \geq p(a_1 = 1, b_3 = 1).$$

We want to prove this statement.

(a) If the source is described by a local hidden-variable theory, then it can be described by a joint probability distribution $P(a_1, a_2, a_3; b_1, b_2, b_3)$. Using the constraint on the source that measurements of corresponding observables must be anticorrelated show that we must have that

$$P(a_1, a_2, a_3; b_1, b_2, b_3) = 0,$$

unless $a_j = -b_j$, $j = 1,2,3$.

(b) Use your result in part (a) to prove Wigner's inequality.

(c) We now want to show that quantum mechanics can satisfy the constraint yet violate the inequality. We choose

$$a_1 = \frac{1}{2}\sigma_z + \frac{\sqrt{3}}{2}\sigma_x \qquad a_2 = \sigma_z \qquad a_3 = \frac{1}{2}\sigma_z - \frac{\sqrt{3}}{2}\sigma_x,$$

and similarly for b_j. Here, $\sigma_x|0\rangle = |1\rangle$ and $\sigma_x|1\rangle = |0\rangle$. The source produces particles in the singlet state

$$|\Phi_-\rangle = \frac{1}{\sqrt{2}}(|0\rangle|1\rangle - |1\rangle|0\rangle),$$

and one of these particles goes to Alice and the other to Bob. This selection of observables and source satisfies the constraint. Show that they do not satisfy Wigner's inequality.

2. Teleportation does not just work for qubits; it works for a quantum system of any dimension. Suppose we want to teleport an N-dimensional quantum state. Let $\{|j\rangle | j = 0, \dots, N-1\}$ be an orthonormal basis for our N-dimensional space, and define the states

$$|\chi_{n,m}\rangle = \frac{1}{\sqrt{N}} \sum_{j=0}^{N-1} e^{2\pi i jn/N} |j\rangle|j+m \ (\text{mod } N)\rangle,$$

for two N-dimensional quantum systems. These will take the place of the Bell states for our more general teleportation procedure.

(a) Show that $\langle \chi_{n,m} | \chi_{n',m'} \rangle = \delta_{n,n'} \delta_{m,m'}$.
(b) We start with the state $|\phi\rangle_{A'} |\chi_{0,0}\rangle_{AB}$, where

$$|\phi\rangle = \sum_{j=0}^{N-1} \alpha_j |j\rangle,$$

is the state we want to teleport. Show that by measuring the $A'A$ particles in the $|\chi_{n,m}\rangle$ basis and using the result of this measurement to perform the correct unitary transformation on the B particle, we can transfer the state $|\phi\rangle$ onto the B particle.

3. We want to consider one step of an entanglement concentration procedure due to A. Sanpera and C. Macchiavello. Start with two pairs of particles, where each pair is in the state consisting of the mixture of qubit Bell states

$$\rho = p|\Psi_+\rangle\langle\Psi_+| + (1-p)|\Psi_-\rangle\langle\Psi_-|,$$

where $p > 1/2$ and $|\Psi_\pm\rangle = (|00\rangle \pm |11\rangle)/\sqrt{2}$. Let us call the pairs AB and $A'B'$, where Alice has particles A and A' and Bob has particles B and B'. Alice applies $\exp(i\pi\sigma_x/4)$ to each of her particles, and Bob applies $\exp(-i\pi\sigma_x/4)$ to both of his. Alice then sends both of her particles into a C-NOT gate, where A is the control qubit and A' is the target, and Bob sends his into a C-NOT gate with B

as the control qubit and B' as the target. Alice and Bob now measure particles A' and B' in the $\{|0\rangle, |1\rangle\}$ basis and keep the pair AB if their results agree. Show that if their results agree, then the proportion of the Bell state $|\Psi_+\rangle$ in the pair AB has increased, so that the pair has become more entangled.

4. Consider the following two-qubit density matrix

$$\rho_{AB} = p|\Phi_-\rangle_{AB \, AB}\langle\Phi_-| + \frac{1-p}{4}I,$$

where $|\Phi_-\rangle_{AB} = (|0\rangle_A|1\rangle_B - |1\rangle_A|0\rangle_B)/\sqrt{2}$.

(a) Use the positive partial transpose condition to find out for what values of p this density matrix is entangled.

(b) Find the concurrence of this density matrix as a function of p.

5. A general bipartite qubit state can be written in the standard basis as $|\psi\rangle_{AB} = a_{00}|00\rangle + a_{01}|01\rangle + a_{10}|10\rangle + a_{11}|11\rangle$. The coefficients can be arranged to form a matrix in a natural way

$$A = \begin{pmatrix} a_{00} & a_{01} \\ a_{10} & a_{11} \end{pmatrix}.$$

Show that $C(|\psi\rangle_{AB}) = 2|\det A|$, where $\det A$ is the determinant of the matrix A and $|\dots|$ is the absolute value of the quantity inside.

6. Consider the two-qubit state

$$\rho = p|\psi\rangle\langle\psi| + \frac{(1-p)}{4}I,$$

where $|\psi\rangle = a|01\rangle + b|10\rangle$. Using the partial transpose condition, find the values of p for which the partial transpose of ρ will have a negative eigenvalue. For p in that range, use the eigenvector corresponding to the negative eigenvalue to construct an entanglement witness for ρ. Express the entanglement witness as a 4×4 matrix in the computational basis.

7. Define the two-mode state $(1/\sqrt{2})(a_1^\dagger + a_2^\dagger)|0\rangle$. Consider the mixed state

$$\rho = p|\psi_{01}\rangle\langle\psi_{01}| + \frac{1-p}{4}P_{01},$$

where $0 \le p \le 1$ and P_{01} is the projection operator onto the space spanned by the vectors $\{|0\rangle_1|0\rangle_2, |0\rangle_1|1\rangle_1, |1\rangle_1|0\rangle_1, |1\rangle_1|1\rangle_2\}$. Using the entanglement condition

$$|\langle a_1 a_2^\dagger\rangle|^2 > \langle a_1^\dagger a_1 a_2^\dagger a_2\rangle,$$

find a range of p for which ρ is definitely entangled.

References

1. M. Hillery, B. Yurke, Bell's theorem and beyond. Quantum Semiclassical Opt. **7**, 215 (1995)
2. C.H. Bennett, G. Brassard, C. Crepeau, R. Jozsa, A. Peres, W. Wootters, Teleporting an unknown quantum state via dual classical and EPR channels. Phys. Rev. Lett. **70**, 1895 (1993)
3. C.H. Bennett, S.J. Wiesner, Communication via one- and two-particle operators on Einstein-Podolsky-Rosen states. Phys. Rev. Lett. **69**, 2881 (1992)
4. C.H. Bennett, D.P. DiVincenzo, J.A. Smolin, W.K. Wootters, Mixed-state entanglement and quantum error correction. Phys. Rev. A **54**, 3824 (1996)
5. C.H. Bennett, H.J. Bernstein, S. Popescu, B. Schumacher, Concentrating partial entanglement by local operations. Phys. Rev. A 53, 2046 (1996)
6. R. Simon, Peres-Horodecki separability criterion for continuous variable states. Phys. Rev. Lett. **84**, 2726 (2000)
7. L.-M Duan, G. Giedke, J.I. Cirac, P. Zoller, Inseparability criterion for continuous variable systems. Phys. Rev. Lett. **84**, 2722 (2000)
8. M. Hillery, M.S. Zubairy, Entanglement conditions for two-mode states. Phys. Rev. Lett. **96**, 050503 (2006)
9. W.K. Wootters, Entanglement of formation of an arbitrary state of two qubits. Phys. Rev. Lett. **80**, 2245 (1998)
10. R. Horodecki, P. Horodecki, M. Horodecki, K. Horodecki, Quantum entanglement. Rev. Mod. Phys. **81**, 865 (2009)
11. O. Gühne, G. Toth, Entanglement detection. Physics Reports **474**, 1 (2009)

Chapter 4
Generalized Quantum Dynamics

Time evolution in textbook quantum mechanics is represented by unitary maps $|\psi\rangle \to U|\psi\rangle$ and $\rho \to U\rho U^\dagger$, where $U = e^{-itH}$. This is not the most general evolution possible. We can couple our system to another one, evolve both with a unitary operator that will, in general, create entanglement between the two systems, and then trace out the second system. The resulting evolution for the original system alone will be non-unitary, in general, and can be described by a non-unitary quantum map.

4.1 Quantum Maps or Superoperators

4.1.1 Quantum Maps and Their Kraus Representation

Let us introduce the unitary map $|\psi\rangle_A \otimes |\psi\rangle_B \to U_{AB}(|\psi\rangle_A \otimes |\psi\rangle_B)$ where $|\psi\rangle_A \otimes |\psi\rangle_B \in \mathcal{H}_A \otimes \mathcal{H}_B$. We apply the identity operator $I_A \otimes I_B$ to the unitary map, where $I_B = \Sigma_m |m\rangle_{BB}\langle m|$ and $\{|m\rangle_B\}$ is an orthonormal basis for $\mathcal{H}_B$, yielding

$$U_{AB}(|\psi\rangle_A \otimes |\psi\rangle_B) = I_A \otimes \Sigma_m |m\rangle_{BB}\langle m|(U_{AB}|\psi\rangle_A \otimes |\psi\rangle_B). \tag{4.1}$$

We can introduce a shorthand for the expression appearing in the right-hand side:

$$_B\langle m|U_{AB}(|\psi\rangle_A \otimes |\psi\rangle_B) \in \mathcal{H}_A \equiv A_m|\psi\rangle_A, \tag{4.2}$$

which defines the operator A_m. In terms of A_m, Eq. (4.1) can be written as

$$U_{AB}(|\psi\rangle_A \otimes |\psi\rangle_B) = \Sigma_m A_m|\psi\rangle_A \otimes |m\rangle_B, \tag{4.3}$$

J.A. Bergou and M. Hillery, *Introduction to the Theory of Quantum Information Processing*, Graduate Texts in Physics, DOI 10.1007/978-1-4614-7092-2_4, © Springer Science+Business Media New York 2013

where $\Sigma_m A_m^\dagger A_m = I_A$ as can be easily seen from its definition. We have just obtained the following mapping:

$$\begin{aligned}
\rho_A &= |\psi\rangle_{AA}\langle\psi| \\
&\to \text{Tr}_B\{\Sigma_m\Sigma_{m'}A_m|\psi\rangle_A \otimes |m\rangle_{BB}\langle m'| \otimes {}_A\langle\psi|A_{m'}^\dagger\} \\
&= \Sigma_m A_m\rho A_m^\dagger.
\end{aligned} \tag{4.4}$$

This gives us what is called the operator sum, or Kraus representation of the quantum map T (or superoperator T),

$$T(\rho) = \Sigma_m A_m\rho A_m^\dagger. \tag{4.5}$$

In the next section we will present a systematic study of the most important properties of quantum maps.

4.1.2 Properties of Quantum Maps

Note that T has a number of important and useful properties.

1. T maps hermitian operators to hermitian operators.
2. T is trace preserving.
3. T maps positive operators to positive operators.

These properties follow directly from the definition of T.

We defined the quantum map and the corresponding Kraus representation as the remainder of a unitary map, defined on a larger Hilbert space, after tracing out part of the system. The converse is also true. Given a Kraus representation, it is possible to find a larger Hilbert space, $\mathcal{H}_A \otimes \mathcal{H}_B$, a vector $|\phi\rangle_B \in \mathcal{H}_B$, and a unitary operator U_{AB} such that

$$A_m|\psi\rangle_A = {}_B\langle m|U_{AB}(|\psi\rangle_A \otimes |\phi\rangle_B). \tag{4.6}$$

We now prove this statement constructively. Let us choose $\mathcal{H}_A$ to have dimension N and $\mathcal{H}_B$ to have dimension M. Further, let $\{|m_B\rangle\}$ be an orthonormal basis for $\mathcal{H}_B$, and choose $|\phi_B\rangle$ to be an arbitrary state of $\mathcal{H}_B$. Let us then define a transformation U_{AB} via

$$U_{AB}(|\psi\rangle_A \otimes |\phi\rangle_B) = \sum_m A_m|\psi\rangle_A \otimes |m\rangle_B, \tag{4.7}$$

which implies the Eq. (4.6). U_{AB} is inner product preserving,

$$\left(\sum_{m'}{}_A\langle\psi'| \otimes {}_B\langle m'|A_{m'}^\dagger\right)\left(\sum_m A_m|\psi\rangle_A \otimes |m\rangle_B\right) = \sum_m {}_A\langle\psi'|A_{m'}^\dagger A_m|\psi\rangle_A$$

$$= {}_A\langle\psi'|\psi\rangle_A, \tag{4.8}$$

so it is unitary on the one-dimensional subspace spanned by $|\phi_B\rangle$ and it can be extended to a full unitary operator on $\mathcal{H}_A \otimes \mathcal{H}_B$ because, e.g., on the subspace that is orthogonal to $|\phi_B\rangle$, it can be the identity.

The Kraus representation of a superoperator is not unique. Let $\{|m_B'\rangle\}$ be a different orthonormal basis for $\mathcal{H}_B$, and

$$B_{m'} = \langle m_B' | U_{AB}(|\psi_A\rangle \otimes |\phi_B\rangle), \tag{4.9}$$

and we have

$$T(\rho_A) = \Sigma_{m'} B_{m'} \rho_A B_{m'}^\dagger. \tag{4.10}$$

If $|m_B'\rangle = \Sigma_m U_{m'm} |m_B\rangle$ then $\langle m_B'| = \Sigma_m U_{mm'}^\dagger \langle m_B|$ and we have that $B_{m'} = \Sigma_m U_{mm'}^\dagger A_m$. We shall eventually show that any two Kraus representations for the same superoperator are related in this way.

First we want to show that a superoperator satisfying conditions 1–3 has a Kraus representation. Actually, we have to replace 3 by a stronger condition:

3'. T is completely positive.

This means the following. We know that T maps bounded operators to bounded operators, T: bounded operators on $\mathcal{H}_A \to$ bounded operators on $\mathcal{H}_A$. Let us append an ancilla space $\mathcal{H}_B$ to $\mathcal{H}_A$ and extend T to a bounded operator on this larger space, so it is in the set of bounded operators $\mathcal{B}(\mathcal{H}_A \otimes \mathcal{H}_B)$, by $T \to T \otimes I_B$. If, for any $\mathcal{H}_B$, $T \otimes I_B$ is positive, then we say that T is completely positive.

Physically what this means is the following. T describes the evolution of the system A, and system B does not evolve. T is completely positive if $\rho_A \otimes \rho_B \to T(\rho_A) \otimes \rho_B$ is a density matrix for any ρ_A and ρ_B. An example of a map that is positive, but not completely positive, is the transpose. It preserves eigenvalues, so it is positive. On the other hand $(\rho)_A^T \otimes I_B$ is just a partial transpose and we have seen that the partial transpose is not positive.

We now want to prove that a superoperator satisfying 1–3' has a Kraus representation. Before proceeding, we note the following method that we will use in the proof. Let A be an operator on $\mathcal{H}_A$, where $\dim \mathcal{H}_A = N$. Suppose $\dim \mathcal{H}_B \geq N$ and let $\{|j_A\rangle\}$ and $\{|j_B\rangle\}$ be the orthonormal bases of $\mathcal{H}_A$ and $\mathcal{H}_B$, respectively. Consider the state

$$|\psi_{AB}\rangle = \Sigma_{j=1}^N \frac{1}{\sqrt{N}} |j_A\rangle \otimes |j_B\rangle. \tag{4.11}$$

If $|\phi_A\rangle \in \mathcal{H}_A$, we can express it in terms of $|\psi_{AB}\rangle$ as a "partial inner product" with a vector $|\phi_B^*\rangle \in \mathcal{H}_B$, where $|\phi_A\rangle = \Sigma_{j=1}^N c_j |j_A\rangle$ and $|\phi_B^*\rangle = \Sigma_{j=1}^N c_j^* |j_B\rangle$. Then

$$\langle \phi_B^* | \psi_{AB} \rangle = (\Sigma_{j=1}^N c_j \langle j_B|) \Sigma_{j'=1}^N \frac{1}{\sqrt{N}} |j_A'\rangle \otimes |j_B'\rangle = \frac{1}{\sqrt{N}} |\phi_A\rangle. \tag{4.12}$$

The mapping $|\phi_A\rangle \to |\phi_B^*\rangle$ is antilinear and norm preserving. We can calculate the effect of $A \otimes I_B$ on $|\psi_{AB}\rangle$ similarly,

$$\langle \phi_B^*|(A \otimes I_B)|\psi_{AB}\rangle = (\Sigma_{j=1}^N c_j \langle j_B|)\Sigma_{j'=1}^N \frac{1}{\sqrt{N}}A|j_A'\rangle \otimes |j_B'\rangle$$

$$= \frac{1}{\sqrt{N}}A(\Sigma_{j=1}^N c_j|j_A\rangle) = \frac{1}{\sqrt{N}}A|\phi_A\rangle. \qquad (4.13)$$

Equipped with this method we can now proceed with the proof. Suppose T is a superoperator satisfying 1, 2, and 3'. T acts on $\mathcal{B}(\mathcal{H}_A)$. $T \otimes I_B$ acting on $\mathcal{B}(\mathcal{H}_A) \otimes \mathcal{B}(\mathcal{H}_B)$ is positive. This implies that if $\rho_{AB} = |\psi_{AB}\rangle\langle\psi_{AB}|$ and

$$\rho_{AB}' = (T \otimes I_B)(\rho_{AB}), \qquad (4.14)$$

then ρ_{AB}' is also a density matrix. It can be expanded as an ensemble of pure states $\rho_{AB}' = \Sigma_\mu q_\mu|\phi_{AB,\mu}\rangle\langle\phi_{AB,\mu}|$. By a derivation entirely similar to those in Eqs. (4.12) and (4.13), we obtain

$$T(|\phi_A\rangle\langle\phi_A|) = N\langle\phi_B^*|(T \otimes I_B)(\rho_{AB})|\phi_B^*\rangle$$

$$= N\Sigma_\mu q_\mu\langle\phi_B^*|\phi_{AB,\mu}\rangle\langle\phi_{AB,\mu}|\phi_B^*\rangle. \qquad (4.15)$$

Now define $A_\mu : |\phi_A\rangle \to \sqrt{Nq_\mu}\langle\phi_B^*|\phi_{AB,\mu}\rangle$. A_μ is a linear operator on $\mathcal{H}_A$, and we have

$$T(|\phi_A\rangle\langle\phi_A|) = \Sigma_\mu A_\mu|\phi_A\rangle\langle\phi_A|A_\mu^\dagger, \qquad (4.16)$$

which we can extend for any density matrix ρ_A to

$$T(\rho_A) = \Sigma_\mu A_\mu\rho_A A_\mu^\dagger. \qquad (4.17)$$

Because T is trace preserving for any ρ_A, we have that $\Sigma_\mu \text{Tr}(\rho_A A_\mu^\dagger A_\mu) = 1$ from which $\Sigma_\mu A_\mu^\dagger A_\mu = I$ follows.

We actually need to show that

$$T(\mathcal{M}_A) = \Sigma_\mu A_\mu \mathcal{M}_A A_\mu^\dagger \qquad (4.18)$$

for any $\mathcal{M}_A \in \mathcal{B}(\mathcal{H}_A)$. We can do this by showing it is true for a basis of $\mathcal{B}(\mathcal{H}_A)$. Such an operator basis is given by $\{(|j_A\rangle\langle k_A|) \mid j,k = 1,\ldots,N\}$. We know the above equation is true for any operator of the form $\mathcal{M}_A = \Sigma_n c_n|\phi_{A,n}\rangle\langle\phi_{A,n}|$. Defining

$$|\phi_{A,1}\rangle = \frac{1}{\sqrt{2}}(|j_A\rangle + |k_A\rangle) \quad |\phi_{A,3}\rangle = \frac{1}{\sqrt{2}}(|j_A\rangle + i|k_A\rangle)$$

$$|\phi_{A,2}\rangle = \frac{1}{\sqrt{2}}(|j_A\rangle - |k_A\rangle) \quad |\phi_{A,4}\rangle = \frac{1}{\sqrt{2}}(|j_A\rangle - i|k_A\rangle) \qquad (4.19)$$

we find

$$|j_A\rangle\langle k_A| = \frac{1}{2}(|\phi_{A,1}\rangle\langle\phi_{A,1}| - |\phi_{A,2}\rangle\langle\phi_{A,2}|) + \frac{i}{2}(|\phi_{A,3}\rangle\langle\phi_{A,3}| - |\phi_{A,4}\rangle\langle\phi_{A,4}|). \quad (4.20)$$

From here it follows immediately

$$T(|j_A\rangle\langle k_A|) = \Sigma_\mu A_\mu |j_A\rangle\langle k_A| A_\mu^\dagger. \quad (4.21)$$

This completes the proof.

We now want to use the construction of the operators A_μ to make some statements about their properties.

4.1.3 Properties of the Kraus Operators

The first question we address is: How many Kraus operators do we need? The mapping in Eq. (4.14) maps $\mathcal{H}_A \otimes \mathcal{H}_B$, where $\mathcal{H}_B = \text{span}\{|j_B\rangle\}$, into itself, and this space has dimension N^2. Diagonalizing ρ'_{AB} we will find at most N^2 vectors in the expansion of ρ'_{AB}. Therefore, there is a Kraus representation with at most N^2 operators.

The next question we address is the uniqueness of the Kraus representation. It is clear that the Kraus representation is not unique, because the decomposition of ρ'_{AB} is not unique. We want to see how different Kraus representations of the same superoperator are related. The idea here is to show that each Kraus representation is related to a decomposition of ρ'_{AB} and then use the theorem about different decompositions of density matrices.

If we have that for any $\mathcal{M}_A \in \mathcal{B}(\mathcal{H}_A)$

$$T(\mathcal{M}_A) = \Sigma_\mu A_\mu \mathcal{M}_A A_\mu^\dagger, \quad (4.22)$$

then

$$(T \otimes I_B)(|\psi_{AB}\rangle\langle\psi_{AB}|) = (T \otimes I_B)\left(\frac{1}{N}\Sigma_{j,j'=1}^N |j_A\rangle \otimes |j_B\rangle\langle j'_A| \otimes \langle j'_B|\right)$$

$$= \frac{1}{N}\Sigma_\mu\Sigma_{j,j'} A_\mu |j_A\rangle \otimes |j_B\rangle\langle j'_A| A_\mu^\dagger \otimes \langle j'_B|. \quad (4.23)$$

Define $\sqrt{q_\mu}|\phi_{AB,\mu}\rangle = \frac{1}{\sqrt{N}}\Sigma_j A_\mu |j_A\rangle \otimes |j_B\rangle$ where $\||\phi_{AB,\mu}\| = 1$, so then

$$\rho'_{AB} = \Sigma_\mu q_\mu |\phi_{AB,\mu}\rangle\langle\phi_{AB,\mu}|. \quad (4.24)$$

Now suppose we have two different Kraus representations for T, $T(\mathcal{M}_A) = \Sigma_\mu A_\mu \mathcal{M}_A A_\mu^\dagger$ and $T(\mathcal{M}_A) = \Sigma_\mu D_\mu \mathcal{M}_A D_\mu^\dagger$. These each give us a decomposition of ρ'_{AB}, $\rho'_{AB} = \Sigma_\mu q_\mu |\phi_{AB,\mu}\rangle\langle\phi_{AB,\mu}| = \Sigma_\nu q'_\nu |\phi'_{AB,\nu}\rangle\langle\phi'_{AB,\nu}|$, where $\sqrt{q_\mu}|\phi_{AB,\mu}\rangle = \frac{1}{\sqrt{N}}\Sigma_j A_\mu |j_A\rangle \otimes |j_B\rangle$ and $\sqrt{q'_\nu}|\phi'_{AB,\nu}\rangle = \frac{1}{\sqrt{N}}\Sigma_j D_\mu |j_A\rangle \otimes |j_B\rangle$.

We know that there exists a unitary matrix $U_{\nu\mu}$ such that

$$\sqrt{q'_\nu}|\phi'_{AB,\nu}\rangle = \Sigma_\mu U_{\nu\mu}\sqrt{q_\mu}|\phi_{AB,\mu}\rangle, \qquad (4.25)$$

or

$$\Sigma_j D_\nu |j_A\rangle \otimes |j_B\rangle = \Sigma_\mu \Sigma_j U_{\nu\mu} A_\mu |j_A\rangle \otimes |j_B\rangle. \qquad (4.26)$$

We can read out from here that

$$D_\nu |j_A\rangle = \Sigma_\mu U_{\nu\mu} A_\mu |j_A\rangle, \qquad (4.27)$$

but $\{|j_A\rangle\}$ is a basis, so, finally,

$$D_\nu = \Sigma_\mu U_{\nu\mu} A_\mu. \qquad (4.28)$$

From here, we can conclude that any two Kraus representations of the same superoperator are related by a unitary matrix.

4.2 An Example: The Depolarizing Channel

We shall now look at an example of a superoperator on qubits, the depolarizing channel. A quantum channel is, in general, a quantum map that maps density matrixes to density matrixes. The idea of the depolarizing channel is that qubit has probability of $(1-p)$ of nothing happening, $\frac{p}{3}$ of σ_x acting on it (bit flip), $\frac{p}{3}$ of σ_z acting on it (phase flip), and $\frac{p}{3}$ of σ_y acting on it (both). One way to do this is to tensor our qubit Hilbert space with a four-dimensional "environment" Hilbert space, $\mathcal{H}_A \otimes \mathcal{H}_E$. We have a unitary operator acting on this tensor product Hilbert space

$$U_{AE}|\psi_A\rangle \otimes |0_E\rangle = \sqrt{1-p}|\psi_A\rangle \otimes |0_E\rangle + \sqrt{\frac{p}{3}}\left(\sigma_x|\psi_A\rangle \otimes |1_E\rangle\right.$$
$$\left. + \sigma_y|\psi_A\rangle \otimes |2_E\rangle + \sigma_z|\psi_A\rangle \otimes |3_E\rangle\right), \qquad (4.29)$$

and, after tracing out the environment we get

$$T(|\psi_A\rangle\langle\psi_A|) = \mathrm{Tr}_E(U_{AE}|\psi_A\rangle \otimes |0_E\rangle\langle\psi_A| \otimes \langle 0_E|U_{AE}^{-1})$$
$$= (1-p)|\psi_A\rangle\langle\psi_A| + \frac{p}{3}\sigma_x|\psi_A\rangle\langle\psi_A|\sigma_x$$
$$+ \frac{p}{3}\sigma_y|\psi_A\rangle\langle\psi_A|\sigma_y + \frac{p}{3}\sigma_z|\psi_A\rangle\langle\psi_A|\sigma_z. \qquad (4.30)$$

From here we can read out the Kraus operators: $A_0 = \sqrt{1-p}I$, $A_1 = \sqrt{\frac{p}{3}}\sigma_x$, $A_2 = \sqrt{\frac{p}{3}}\sigma_y$, and $A_3 = \sqrt{\frac{p}{3}}\sigma_z$. It is easy to check that $\Sigma_{\mu=0}^{3}A_\mu^\dagger A_\mu = I$, which is a direct consequence of the unitarity of U_{AE}.

It is interesting to see what happens to the Bloch sphere under this mapping. Let

$$\rho = \frac{1}{2}(I + \mathbf{n} \cdot \sigma), \tag{4.31}$$

and use

$$\sigma_j\sigma_k\sigma_j = \begin{cases} -\sigma_k \ k \neq j \\ \sigma_j \ j = k \end{cases}. \tag{4.32}$$

Then

$$T(\sigma_x) = (1-p)\sigma_x + \frac{p}{3}\sigma_x - \frac{2p}{3}\sigma_x = \left(1 - \frac{4p}{3}\right)\sigma_x, \tag{4.33}$$

and, similarly, $T(\sigma_y) = (1 - \frac{4p}{3})\sigma_y$ and $T(\sigma_z) = (1 - \frac{4p}{3})\sigma_z$. So, finally

$$T(\rho) = \frac{1}{2}(I + \mathbf{n}' \cdot \sigma), \tag{4.34}$$

where $\mathbf{n}' = (1 - \frac{4p}{3})\mathbf{n}$.

This tells us that the map representing the depolarizing channel just causes the entire Bloch sphere to contract by a factor of $|1 - (4p/3)|$.

4.3 Impossible Maps

4.3.1 The Cloning Map and the No-Cloning Theorem

It is also useful to know that certain maps are impossible. One of them is the "cloning" map that would duplicate quantum states. Suppose we have a device that does copy qubit quantum states. A general input for such a device is of the form $|\psi_a\rangle \otimes |0_b\rangle \otimes |Q_c\rangle$ where $|\psi_a\rangle$ is the state of qubit a to be copied, $|0_b\rangle$ is a blank initial state of qubit b that becomes the copy, and $|Q_c\rangle$ is an ancillary state which can be interpreted as the initial state of the copier. What we want is the unitary map

$$U(|\psi_a\rangle \otimes |0_b\rangle \otimes |Q_c\rangle) = |\psi_a\rangle \otimes |\psi_b\rangle \otimes |Q_{\psi,c}\rangle \tag{4.35}$$

Since the copier must work for arbitrary states, so, in particular, it must copy basis states

$$U(|0_a\rangle \otimes |0_b\rangle \otimes |Q_c\rangle) = |0_a\rangle \otimes |0_b\rangle \otimes |Q_{0,c}\rangle,$$
$$U(|1_a\rangle \otimes |0_b\rangle \otimes |Q_c\rangle) = |1_a\rangle \otimes |1_b\rangle \otimes |Q_{1,c}\rangle. \tag{4.36}$$

These relations determine how a general state is copied. If $|\psi\rangle = \alpha|0\rangle + \beta|0\rangle$, then multiplying the first equation by α and the second by β and adding them together gives

$$U(|\psi_a\rangle \otimes |0_b\rangle \otimes |Q_c\rangle) = \alpha|0_a\rangle \otimes |0_b\rangle \otimes |Q_{0,c}\rangle + \beta|1_a\rangle \otimes |1_b\rangle \otimes |Q_{1,c}\rangle, \quad (4.37)$$

and this is not the same as Eq. (4.35). This is known as the no-cloning theorem. It is a direct consequence of the linearity of quantum mechanics and implies that quantum information is very different from classical information, which can be copied. The fact that quantum information cannot be copied can also be used to our advantage. In particular, it is one of the reasons why quantum cryptography works.

4.3.2 Faster than Light Communication

If cloning were possible, superluminal communication would be, too. To demonstrate this, suppose that Alice and Bob initially share an ebit, $|\psi_{AB}\rangle = \frac{1}{\sqrt{2}}(|0_A\rangle|1_B\rangle - |1_A\rangle|0_B\rangle)$. In the $|\pm\rangle = \frac{1}{\sqrt{2}}(|0\rangle \pm |1\rangle)$ basis, the same state can be written as $|\psi_{AB}\rangle = -\frac{1}{\sqrt{2}}(|+_A\rangle|-_B\rangle - |-_A\rangle|+_B\rangle)$. Alice now measures her particle either in the $\{|0\rangle, |1\rangle\}$ basis or in the $\{|+\rangle, |-\rangle\}$ basis, while Bob clones his particle, making $2N$ copies. He then measures N in the $\{|0\rangle, |1\rangle\}$ basis and N in the $\{|+\rangle, |-\rangle\}$ basis. The basis in which all measurements produce the same results tells Bob which basis Alice measured in.

4.4 Problems

1. We want to show that a superoperator is invertible if and only if it is unitary, i.e., $M(B) = UBU^\dagger$, for any $B \in \mathcal{B}(\mathcal{H})$. It is clear that if M is unitary, then it is invertible. We need to now show the opposite, i.e., that if M is invertible, then it is unitary. Let

$$M(|\psi\rangle\langle\psi|) = \sum_\mu M_\mu|\psi\rangle\langle\psi|M_\mu^\dagger.$$

The superoperator N is the inverse of M if $N \circ M = I$, or

$$\sum_{\mu,\nu} N_\nu M_\mu |\psi\rangle\langle\psi|M_\mu^\dagger N_\nu^\dagger = |\psi\rangle\langle\psi|,$$

for all $|\psi\rangle$.

(a) Use the fact that

$$\sum_{\mu,\nu} |\langle \psi | N_\nu M_\mu | \psi \rangle|^2 = 1,$$

which is implied by the above equation, and the normalization conditions on the operators M_μ and N_ν to show that $N_\nu M_\mu = \lambda_{\nu\mu} I$.

(b) Use the result in part (a) to show that $M_{\mu'}^\dagger M_\mu$ is a multiple of the identity for any μ and μ'.

(c) Use the result in part (b) to show that M is unitary.

2. (a) Let $|\psi_1\rangle$ and $|\psi_2\rangle$ be two one-qubit states. Find a value of ϕ for which the map

$$|\psi_1\rangle |\psi_2\rangle \to \frac{1}{\sqrt{2}} (|\psi_2\rangle |\psi_1\rangle + e^{i\phi} |\psi_1\rangle |\psi_2\rangle),$$

can be realized as a unitary operation.

(b) The above transformation can be used to spread the information in a single qubit over two qubits. Suppose $|\psi_1\rangle = |0\rangle$ and $|\psi_2\rangle = |\psi\rangle$. Now suppose we lose one of the qubits. Find the reduced density matrix of the remaining qubit and its fidelity to the state $|\psi\rangle$. By spreading the information contained in one qubit over two, we retain some information about the original qubit even though one of the qubits is lost.

(c) For a general one-qubit state $|\psi\rangle$, find a value of ϕ so that the transformation

$$|0\rangle |0\rangle |\psi\rangle \to \frac{1}{\sqrt{3}} [|0\rangle |0\rangle |\psi\rangle + e^{i\phi} (|0\rangle |\psi\rangle |0\rangle + |\psi\rangle |0\rangle |0\rangle)].$$

can be realized by a unitary operator.

3. A C-NOT gate is a rather versatile device that can be used to realize a number of maps.

(a) Suppose the input state is $(\sqrt{p_0} |0\rangle + \sqrt{p_1} |1\rangle) \otimes |\psi\rangle$, where $p_0 + p_1 = 1$, $|\psi\rangle$ is a general one-qubit state, and the first qubit is the control qubit, and the second is the target qubit. We send this state through the C-NOT gate and trace out the control qubit. Find the Kraus operators for the resulting map on the target qubit.

(b) Now consider the input state $|\psi\rangle \otimes (1/\sqrt{2})(e^{i\theta} |0\rangle + e^{-i\theta} |1\rangle)$. Send this state through the C-NOT and trace out the target qubit. Find the Kraus operators for the resulting map on the control qubit.

4. The SWAP operator on two qubits acts as $S|\psi\rangle_a \otimes |\phi\rangle_b = |\phi\rangle_a \otimes |\psi\rangle_b$. A partial SWAP operator is given by $P(\theta) = \cos\theta I_{ab} + i\sin\theta S$, where I_{ab} is the identity operator on two qubits. It can be thought of as an operator that partially exchanges the information between two qubits.

(a) Show that if we consider the two-qubit density matrix given by $\rho_a \otimes \xi_b$, where ρ_a and ξ_b are one-qubit density matrices, that

$$\rho_a' = \text{Tr}_b[P(\theta)\rho_a \otimes \xi_b P^\dagger(\theta)],$$

is given by

$$\rho_a' = \cos^2 \theta \rho_a + \sin^2 \theta \xi_a + i\cos \theta \sin \theta [\xi_a, \rho_a].$$

(b) Expressing ρ_a and ρ_a' in Bloch form

$$\rho_a = \frac{1}{2}I_a + \mathbf{r} \cdot \sigma, \quad \rho_a' = \frac{1}{2}I_a + \mathbf{r}' \cdot \sigma,$$

find $\mathbf{r}'$ in terms of $\mathbf{r}$.

References

1. J. Preskill, *Lecture Notes for Physics 219* http://www.theory.caltech.edu/people/preskill/ph229/
2. W.K. Wooters, W.H. Zurek, A single quantum cannot be cloned. Nature **299**, 802 (1982)

Chapter 5
Quantum Measurement Theory

5.1 Outline

Measurements are an integral part of quantum information processing. Reading out the quantum information at the end of the processing pipeline is equivalent to learning what final state the system is in at the output since information is encoded in the state. In fact, information is the state itself. Since finding out the state of a system can be done only by performing measurements on it, we need a thorough understanding of the quantum theory (and practice) of measurements. To this end we will begin with a simplified model of a quantum measurement, due essentially to von Neumann, and from this model we'll read out the postulates of standard quantum measurement theory. Then, by analyzing the underlying assumptions, we'll show that some of the postulates can be replaced by more relaxed ones and this will lead us to the concept of generalized measurements (POVMs) which are particularly useful in measurement optimization problems. Next, by invoking Neumark's theorem we will show how to actually implement positive operator valued measures (POVMs) experimentally. As illustrations of these general concepts we will study two state discrimination strategies in some detail, namely, the unambiguous discrimination and the minimum-error discrimination of two quantum states. As an example for the application of the ideas developed in this chapter we will analyze the B92 quantum key distribution (QKD) protocol in Sect. 6.3. QKD is the crucial ingredient of most quantum cryptographic protocols and in the B92 proposal all of the concepts of this chapter come together in a particularly clean and instructive way.

5.2 Standard Quantum Measurements

We begin with a brief review of a simplified model of a quantum measurement. Let us assume that we want to measure a physical quantity to which, in quantum mechanics, there corresponds a hermitian operator X. The measurement then

J.A. Bergou and M. Hillery, *Introduction to the Theory of Quantum Information Processing*, Graduate Texts in Physics, DOI 10.1007/978-1-4614-7092-2_5, © Springer Science+Business Media New York 2013

consists of the following process. We couple this observable to a so-called pointer variable, the states of which we assume to be macroscopically distinguishable. This is equivalent to assuming that the states of the pointer variable are essentially classical and it is our basic assumption that classical states can be readily measured. For example, our pointer can be a freely propagating heavy particle, and the pointer variable that we observe is simply its position. The initial state of the pointer is a narrow but not too narrow wave packet. What we mean by this is the following. On the one hand, the wave packet must be narrow enough so that the possible pointer positions are clearly distinguishable; there is no overlap among them. On the other hand, the wave packet should not be narrower than necessary because if it is it will spread too fast during the time of the measurement. Let us make this a little more quantitative. From the uncertainty principle, $\Delta x \Delta p \simeq \hbar$, we obtain $\Delta p \simeq \hbar/\Delta x$ which leads to an uncertainty $\Delta v = \hbar/m\Delta x$ in the speed of the pointer particle. Thus, during the time of the measurement, the spread of the initial wave packet increases as

$$\Delta x(t) = \Delta x + \frac{\hbar t}{m \Delta x}. \tag{5.1}$$

This expression is a minimum for a given measurement time t if the spread of the initial wave packet is $\Delta x_{\mathrm{opt}} = \sqrt{\frac{\hbar t}{m}}$, yielding

$$\Delta x_{\min}(t) \equiv \Delta x_{\mathrm{SQL}} = 2\sqrt{\frac{\hbar t}{m}}, \tag{5.2}$$

where SQL stands for Standard Quantum Limit. The initial wave packet should not be prepared narrower than Δx_{opt} which in most cases is not a serious restriction since m is large.

Next, we introduce a coupling between the system and the pointer. The full Hamiltonian is given by

$$H = H_0 + \frac{P^2}{2m} + \hbar g X P, \tag{5.3}$$

where, on the right-hand side, H_0 is the Hamiltonian of the system, the next term is the kinetic energy of the pointer, and the last term is the coupling between the system and the pointer, g being the coupling constant. Since we want to observe the position of the pointer, we choose the coupling between the complementary quantity, the canonical momentum P of the pointer, and the observable X of the system that we want to measure. For simplicity, we assume that the observable X commutes with the unperturbed Hamiltonian H_0 of the system. The above Hamiltonian leads to the time evolution described by the unitary operator

$$U(t) = e^{-igtXP}. \tag{5.4}$$

The observable X is hermitian, so it does have a spectral representation, $X = \sum_j \lambda_j P_j = \sum_j \lambda_j |j\rangle\langle j|$ where λ_j are the (real) eigenvalues, $|j\rangle$ the corresponding eigenstates, and $P_j = |j\rangle\langle j|$ the projector on the subspace spanned by $|j\rangle$. The eigenstates form a complete set in the Hilbert space of the system which is equivalent to saying that the projectors span the identity, $\sum_j P_j = 1$. Using this last relation we can write the time evolution operator as

$$U(t) = \sum_j e^{-ix_j P}|j\rangle\langle j|, \tag{5.5}$$

where we introduced the notation

$$x_j = gt\lambda_j. \tag{5.6}$$

Now, we assume that the joint system-pointer system was initially prepared in the state $\sum_j c_j |j\rangle \otimes |\psi(x)\rangle$ where $|\psi^S\rangle = \sum_j c_j |j\rangle$ is an arbitrary initial state of the system and $|\psi(x)\rangle$ is the initial state of the pointer which we assume to be a well-localized wave packet around $x = 0$, as discussed above. If we apply the time evolution operator, Eq. (5.5), to this initial state, we obtain the joint system—pointer state after the measurement time t:

$$|\psi^{SP}\rangle = \sum_j c_j |j\rangle |\psi(x - x_j)\rangle. \tag{5.7}$$

What we see from here is that there is a very strong correlation between the state of the pointer and the state of the system. We assume that the pointer is essentially classical, so we will always find it in one of the new positions at $x = x_j$. When it is found at $x = x_j$ the state of the system is $c_j |j\rangle$. Since $x_j = gt\lambda_j$ is uniquely related to the eigenvalue λ_j, we can say that by observing the position of the pointer after the measurement we have measured the observable X and found one of its eigenvalues λ_j as the measurement result. Furthermore, the (non-normalized) state of the system, if this particular value was found, is just $c_j |j\rangle$. Taking the inner product of $|\psi^S\rangle = \sum_j c_j |j\rangle$ with $|j\rangle$ gives $c_j = \langle j|\psi^S\rangle$, which tells us that the non-normalized postmeasurement state is $|j\rangle\langle j|\psi^S\rangle = P_j |\psi^S\rangle$. The normalized state after the measurement, if the particular outcome λ_j was found, is $|\phi_j\rangle = \frac{P_j|\psi^S\rangle}{|c_j|}$.

These findings are summarized by the postulates of quantum measurement theory in a more formal way. However, before we list these postulates, we want to have an expression for the resolution of the above measurement. Obviously, we can resolve the different pointer positions if their distance is larger than the SQL, $\Delta x_j = x_{j+1} - x_j = gt\Delta\lambda_j \geq x_{SQL}$. When we use the relation between the pointer position and the eigenvalues, Eq. (5.6), we will find the resolution limit, as

$$\Delta\lambda_j \geq \frac{2}{g}\sqrt{\frac{\hbar}{mt}}, \tag{5.8}$$

which is the minimum separation of the eigenvalues that can be resolved in a quantum measurement. As expected, with increasing measurement time the resolution improves.

We are now in the position to read out the postulates of the quantum measurement theory from the preceding discussion. Let us assume that we are measuring the observable X which has the spectral representation $X = \sum_j \lambda_j |j\rangle\langle j|$. From the hermiticity of X it follows that the eigenvalues λ_j are real. For simplicity we assume that the eigenvalues are nondegenerate and the corresponding eigenvectors, $\{|j\rangle\}$, form a complete orthonormal basis set. Then

1. The projectors $P_j = |j\rangle\langle j|$ span the entire Hilbert space, $\sum_j P_j = 1$.
2. From the orthogonality of the states we have $P_i P_j = P_i \delta_{ij}$. In particular, $P_i^2 = P_i$ from which it follows that the eigenvalues of any projector are 0 and 1.
3. A measurement of X yields one of the eigenvalues λ_j.
4. The state of the system after the measurement is $|\phi_j\rangle = \dfrac{P_j|\psi\rangle}{\sqrt{\langle\psi|P_j|\psi\rangle}}$ if the outcome is λ_j.
5. The probability that this particular outcome is found as the measurement result is $p_j = ||P_j\psi\rangle||^2 = \langle\psi|P_j^2|\psi\rangle = \langle\psi|P_j|\psi\rangle$ where we used the property 2.
6. If we perform the measurement but we do not record the results, the post-measurement state can be described by the density operator $\rho = \sum_j p_j |\phi_j\rangle\langle\phi_j| = \sum_j P_j|\psi\rangle\langle\psi|P_j$.

These six postulates adequately describe what happens to the system during the measurement if it was initially in a pure state. If the system is initially in the mixed state ρ the last three postulates are to be replaced by their immediate generalizations:

4a. The state of the system after the measurement is $\rho_j = \dfrac{P_j \rho P_j}{\mathrm{Tr}(P_j \rho P_j)} = \dfrac{P_j \rho P_j}{\mathrm{Tr}(P_j \rho)}$ if the outcome is λ_j.
5a. The probability that this particular outcome is found as the measurement result is $p_j = Tr(P_j \rho P_j) = Tr(P_j^2 \rho) = Tr(P_j \rho)$ where, again, we used the property 2.
6a. If we perform the measurement but we do not record the results, the post-measurement state can be described by the density operator $\tilde{\rho} = \sum_j p_j \rho_j = \sum_j P_j \rho P_j$.

Of course, 4a–6a reduce to 4–6 for the pure state density matrix $\rho = |\psi\rangle\langle\psi|$. Therefore, in what follows we use the density matrix to describe a general (pure or mixed) quantum state unless we want to emphasize that the state is pure.

Let us summarize the message of these postulates. They essentially tell us that the measurement process is random; we cannot predict its outcome. What we can predict is the spectrum of the possible outcomes and the probability that a particular outcome is found in an actual measurement. This leads us to the ensemble interpretation of quantum mechanics. The state $|\psi\rangle$ (or ρ for mixed states) describes not a single system but an ensemble of identically prepared systems. If we perform the same measurement on each member of the ensemble we can predict the possible measurement results and the probabilities with which they occur, but we cannot predict the outcome of an individual measurement, except, of course, when the probability of a certain outcome is 0 or 1. With the help of these postulates we can then calculate the moments of the probability distribution, $\{p_j\}$, generated by the measurement. The first moment is the average of a large number of identical measurements performed on the initial ensemble. It is called the expectation value of X and is denoted as $\langle X \rangle$,

$$\langle X \rangle = \sum_j \lambda_j p_j = \sum_j \lambda_j Tr(P_j \rho) = Tr(X\rho), \qquad (5.9)$$

where we used the spectral representation of X. The second moment, $\langle X^2 \rangle = Tr(X^2 \rho)$, is related to the variance σ,

$$\sigma^2 = \langle (X - \langle X \rangle)^2 \rangle = \langle X^2 \rangle - \langle X \rangle^2. \qquad (5.10)$$

Higher moments can also be calculated in a straightforward manner, but typically the first and second moments are the most important ones to consider.

5.3 Positive Operator Valued Measures

Now we are in the position to put the postulates of standard measurement theory under closer scrutiny. What the last three postulates provide us with is, in fact, an algorithm to generate probabilities. The generated probabilities are nonnegative, $0 \le p_j \le 1$, and the probability distribution is normalized to unity, $\sum_j p_j = 1$, which is a consequence of the first two postulates. Furthermore, the number of possible outcomes is bounded by the number of terms in the orthogonal decomposition of the identity operator of the Hilbert space. Obviously, one cannot have more orthogonal projections than the dimensionality, N_A, of the Hilbert space of the system, so $j \le N_A$. It would, however, be often desirable to have more outcomes than the dimensionality while keeping the positivity and normalization of the probabilities. We will first show that this is formally possible: if we relax the above rather restrictive postulates and replace them with more flexible ones we can still obtain a meaningful probability generating algorithm. Then we will show that there are physical processes that fit these more general postulates.

Let us begin with the formal considerations and take a closer look at Postulate 5a (or 5) which is the one that gives us the prescription for the generation of probabilities. We notice that in order to get a positive probability with this prescription it is sufficient if P_j^2 is a positive operator; we do not need to require the positivity of an underlying P_j operator. So let us try the following. We introduce a positive operator, $\Pi_j \ge 0$, which is the generalization of P_j^2, and prescribe $p_j = Tr(\Pi_j \rho)$. Of course, we want to ensure that the probability distribution generated by this new prescription is still normalized. Inspecting the postulates we can easily figure out that normalization is a consequence of Postulate 1 and, therefore, requires that $\sum_j \Pi_j = I$, that is, the positive operators still represent a decomposition of the identity. We will call a decomposition of the identity in terms of positive operators, $\sum_j \Pi_j = I$, a POVM, and $\Pi_j \geq 0$ the elements of the POVM. These generalizations will form the core of our new postulates $1'$ and $5'$.

As observed in the previous paragraph, for a POVM to exist, we do not have to require orthogonality and positivity of the underlying P_j operators. Therefore, the underlying operators that, via Postulates 4 (or 4a) and 6 (or 6a), determine the postmeasurement state can be just about any operators, even non-hermitian ones. For projective measurements orthogonality was essentially a consequence of Postulate 2, which was our most constraining postulate because it restricted the number of terms in the decomposition of the identity to at most the dimensionality of the system. Let us now see how far we can get by abandoning it.

If we abandon Postulate 2 then the operators that generate the probability distribution are no longer the same as the ones that generate the postmeasurement states and we have a considerable amount of freedom in choosing them. Let us denote the operators that generate the postmeasurement state by A_j, they are the generalizations of the orthogonal projectors, P_j. In other words, we define the non-normalized postmeasurement state by $A_j|\psi\rangle$ and the corresponding normalized state after the measurement by $|\phi\rangle = A_j|\psi\rangle / \sqrt{\langle\psi|A_j^\dagger A_j|\psi\rangle}$. This expression will form the essence of our new Postulate $4'$. It immediately tells us that Π_j has the structure $\Pi_j = A_j^\dagger A_j$ which by construction is a positive operator. Let us now use our freedom in designing the postmeasurement state. First note that, since the POVM elements are positive operators, $\Pi_j^{1/2}$ exists. Obviously, this is a possible choice for A_j. So is

$$A_j = U_j \Pi_j^{1/2}, \tag{5.11}$$

where U_j is an arbitrary unitary operator. This is the most general form of the detection operators, satisfying $A_j^\dagger A_j = \Pi_j$, and the above expression corresponds to their polar decomposition. What we see is that the POVM elements Π_j determine the absolute value operators through $|A_j| = \Pi_j^{1/2}$ but leave their unitary part open. The A_j operators represent a generalization of the projectors P_j, whereas Π_j is a generalization of P_j^2. The set $\{A_j\}$ is called the set of detection operators and these operators figure prominently in our new postulates $2', 4',$ and $6'$ replacing the corresponding ones of the standard measurements.

With this we completed the goal we set out at the beginning of this section, namely, the generalization of all of the postulates of the standard measurement theory to more flexible ones while keeping the spirit of the old ones. It is now time to list our new postulates.

$1'$. We consider the decomposition of the identity, $\sum_j \Pi_j = 1$, in terms of positive operators, $\Pi_j \geq 0$. Such a decomposition is called a POVM and the Π_j the elements of the POVM.

$2'$. The elements of the POVM, Π_j, can be expressed in terms of the detection operators A_j as $\Pi_j = A_j^\dagger A_j$ where, in general, the detection operators are non-hermitian ones, restricted only by the requirement $\sum_j A_j^\dagger A_j = I$. Then, by construction, the POVM elements are positive operators. Conversely, for a given POVM the detection operators can be expressed in terms of the POVM elements as $A_j = U_j \Pi_j^{1/2}$ where U_j is an arbitrary unitary operator.

$3'$. A detection yields one of the alternatives corresponding to an element of the POVM.

$4'$. The state of the system after the measurement is $|\phi_j\rangle = \dfrac{A_j|\psi\rangle}{\sqrt{\langle\psi|A_j^\dagger A_j|\psi\rangle}}$ if it was initially in the pure state $|\psi\rangle$, and $\rho_j = \dfrac{A_j\rho A_j^\dagger}{Tr(A_j\rho A_j^\dagger)} = \dfrac{A_j\rho A_j^\dagger}{Tr(A_j^\dagger A_j\rho)}$ if it was initially in the mixed state ρ. The inclusion of the arbitrary unitary operator U_j into the detection operator gives us a great deal of flexibility in designing the postmeasurement state.

$5'$. The probability that this particular alternative is found as the measurement result is $p_j = Tr(A_j\rho A_j^\dagger) = Tr(A_j^\dagger A_j\rho) = Tr(\Pi_j\rho)$ where we used the cyclic property of the trace operation.

$6'$. If we perform the measurement but we do not record the results, the postmeasurement state is described by the density operator $\tilde{\rho} = \sum_j p_j\rho_j = \sum_j A_j\rho A_j^\dagger$.

Very often we are not concerned with the state of the system after such operation is performed but only with the resulting probability distribution. For this, it is sufficient to consider Postulates $1'$ and $5'$ defining the probability of finding alternative j as the detection result. Note, that at no step did we require the orthogonality of the Π_j's. Since orthogonality is no longer a requirement, the number of terms in this decomposition of the identity is not bounded by N_A. In fact, the number of terms can be arbitrary. Obviously, what we arrived at is a generalization of the von Neumann projective measurement. It is a surprising generalization as it tells us that just about any operation that satisfies Postulates $1'$ and $2'$ is a legitimate operation that generates a valid probability distribution. It is also a rather natural generalization of the standard quantum measurement since it provides us with a well-defined algorithm that generates a well-behaved probability distribution. So this procedure can be regarded as a *generalized measurement*, and, indeed, for most purposes, it is a sufficient generalization of the standard quantum measurement.

A further note is in place here. Very often a projector projects on a one-dimensional subspace, which is spanned by the vector $|\omega_j\rangle$, in which case it can be written as $P_j = |\omega_j\rangle\langle\omega_j|$. The corresponding generalization to a non-hermitian detection operator can be written as $A_j = c_j|\tilde{\omega}_j\rangle\langle\omega_j|$ where $\langle\omega_j|\omega_j\rangle = \langle\tilde{\omega}_j|\tilde{\omega}_j\rangle = 1$, and c_j is a complex number inside the unit circle $|c_j|^2 \le 1$, and $\langle\omega_j|\tilde{\omega}_j\rangle$ is arbitrary. Then

$$\Pi_j = |c_j|^2|\omega_j\rangle\langle\omega_j|, \tag{5.12}$$

but $\langle\omega_j|\omega_k\rangle \not\propto \delta_{jk}$, and, hence, it is explicit that the POVM is not an orthogonal decomposition of the identity. Since $A_j|\psi\rangle = c_j\langle\omega_j|\psi\rangle|\tilde{\omega}_j\rangle$, we see that

$|\tilde{\omega}_j\rangle$ is proportional to the postmeasurement state $|\phi_j\rangle$, and $p_j = \langle\psi|\Pi_j|\psi\rangle = \langle\psi|A_j^\dagger A_j|\psi\rangle = |c_j|^2|\langle\omega_j|\psi\rangle|^2$. So, we have that

$$|c_j|^2 = \frac{p_j}{|\langle\omega_j|\psi\rangle|^2}, \tag{5.13}$$

and

$$\Pi_j = \frac{p_j}{|\langle\omega_j|\psi\rangle|^2}|\omega_j\rangle\langle\omega_j|. \tag{5.14}$$

Of course, up to this point all this is just a formal mathematical generalization of the standard quantum measurement. The important question is, how can we implement such a thing physically? In the next section we set out to answer this question and then we will study examples of POVMs.

5.4 Neumark's Theorem and the Implementation of a POVM Via Generalized Measurements

First, let us take a look at what happens if we couple our system to another system called ancilla, let them evolve, and then measure the ancilla. The Hilbert space of this larger system is $\mathcal{H}_A \otimes \mathcal{H}_B$, using the tensor product extension, where the Hilbert space of the original system is $\mathcal{H}_A$ and the Hilbert space of the ancilla is $\mathcal{H}_B$. We want to gain information about the state of the system that we now denote as $|\psi_A\rangle$. We assume that the system and the ancilla are initially independent; their joint initial state is $|\psi_A\rangle \otimes |\psi_B\rangle$. Let $\{|m_B\rangle\}$ be an orthonormal basis for $\mathcal{H}_B$ and U_{AB} a unitary operator acting on $\mathcal{H}_A \otimes \mathcal{H}_B$. The probability p_m of measuring $|m_B\rangle$ is then given by

$$p_m = \|(I_A \otimes |m_B\rangle\langle m_B|)U_{AB}(|\psi_A\rangle \otimes |\psi_B\rangle)\|^2. \tag{5.15}$$

Define

$$A_m|\psi_A\rangle \equiv \langle m_B|U_{AB}(|\psi_A\rangle \otimes |\psi_B\rangle). \tag{5.16}$$

Then A_m is a linear operator on $\mathcal{H}_A$ that depends on $|m_B\rangle$, $|\psi_B\rangle$, and U_{AB}. With the help of this definition we can write the measurement probability as

$$p_m = \|A_m|\psi_A\rangle \otimes |m_B\rangle\|^2 = \langle\psi_A|A_m^\dagger A_m|\psi_A\rangle. \tag{5.17}$$

Note that

$$\sum_m \langle\psi_A|A_m^\dagger A_m|\psi_A\rangle = \sum_m ((\langle\psi_A| \otimes \langle\psi_B|)U_{AB}^\dagger|m_B\rangle\langle m_B|U_{AB}(|\psi_A\rangle \otimes |\psi_B\rangle))$$

$$= 1. \tag{5.18}$$

Since this is true for any $|\psi_A\rangle$, we must have that

$$\sum_m A_m^\dagger A_m = I_A, \tag{5.19}$$

where I_A is the identity in $\mathcal{H}_A$.

The non-normalized state of the total "system plus ancilla" after the measurement is $A_m|\psi_A\rangle \otimes |m_B\rangle$, so the (normalized) postmeasurement state of the system alone is

$$|\phi_A\rangle = \frac{1}{\sqrt{\langle\psi_A|A_m^\dagger A_m|\psi_A\rangle}} A_m|\psi_A\rangle. \tag{5.20}$$

Clearly, the (normalized) state of the ancilla after the measurement is $|\phi_B\rangle = |m_B\rangle$, up to an arbitrary phase factor. After the measurement is done and outcome $|m_B\rangle$ is found, the ancilla is no longer of interest and can be discarded.

The set $\{A_m^\dagger A_m\}$ thus gives a decomposition of the identity in terms of positive operators. Therefore, we can identify the set with a POVM where $\{A_m^\dagger A_m\}$ are its elements. In fact, what we see here is the first half of Neumark's theorem: If we couple our system to an ancilla, let them evolve so that they become entangled, and perform a measurement on the ancilla, which collapses the ancilla to one of the basis vectors of the ancilla space, then this procedure will also transform the system because the ancilla degrees of freedom are now entangled to the system. The transformation of the state of the system is, however, neither unitary nor a projection. It can adequately be described as a POVM, so the above procedure corresponds to a POVM in the system Hilbert space. Thus, we have just found a procedure that, when we look at the system only, looks like a POVM. We now know that there are physical processes that can adequately be described as POVMs.

Next we address the question, given the set of operators $\{A_m\}$ acting on $\mathcal{H}_A$ such that $\sum_m A_m^\dagger A_m = I$, can this be interpreted as resulting from a measurement on a larger space? That is, can we find $\mathcal{H} = \mathcal{H}_A \otimes \mathcal{H}_B$, $|\psi_B\rangle$, $\{|m_B\rangle\} \in \mathcal{H}_B$, and U_{AB} acting on $\mathcal{H}$ such that

$$A_m|\psi_A\rangle = \langle m_B|U_{AB}(|\psi_A\rangle \otimes |\psi_B\rangle)) \tag{5.21}$$

holds?

The answer to this question is yes, and we will now prove it constructively. Let us choose $\mathcal{H}_B$ to have dimension M and let $\{|m_B\rangle\}$ be an orthonormal basis for $\mathcal{H}_B$, and choose $|\psi_B\rangle$ to be an arbitrary but fixed initial state in $\mathcal{H}_B$. Let us further define a transformation U_{AB} via

$$U_{AB}(|\psi_A\rangle \otimes |\psi_B\rangle) = \sum_m A_m|\psi_A\rangle \otimes |m_B\rangle, \tag{5.22}$$

which implies the Eq. (5.21). U_{AB} is inner product preserving,

$$\left(\sum_{m'}\langle\psi'_A|A^\dagger_{m'}\otimes\langle m'_B|\right)\left(\sum_m A_m|\psi_A\rangle\otimes|m_B\rangle\right) = \sum_m\langle\psi'_A|A^\dagger_m A_m|\psi_A\rangle$$

$$= \langle\psi'_A|\psi_A\rangle, \qquad (5.23)$$

so it is unitary on the one-dimensional subspace spanned by $|\psi_B\rangle$ and it can be extended to a full unitary operator on $\mathcal{H}_A\otimes\mathcal{H}_B$ because, e.g., on the subspace that is orthogonal to $|\psi_B\rangle$ it can be the identity.

This completes the proof of Neumark's theorem which asserts that there is a one-to-one correspondence between a POVM and the above procedure which sometimes is itself called a generalized measurement. Hence, a generalized measurement can be regarded as the physical implementation of a given POVM.

To close this section we will now illustrate these general considerations on an example. The example is an application of the minimum-error state discrimination strategy that will be discussed in the next section. Suppose one is given a qubit which is prepared equally likely in either of the following three states:

$$|\psi_0\rangle = -\frac{1}{2}\left(|0\rangle + \sqrt{3}\,|1\rangle\right),$$

$$|\psi_1\rangle = -\frac{1}{2}\left(|0\rangle - \sqrt{3}\,|1\rangle\right),$$

$$|\psi_2\rangle = |0\rangle, \qquad (5.24)$$

that is, the probability that $|\psi_j\rangle$ (for $j = 0,1,2$) is prepared is $1/3$. These three states form an overcomplete set of symmetric states that is known as the trine ensemble. For the minimum-error discrimination consider the operators

$$\Pi_j = A^\dagger_j A_j = \frac{2}{3}\,|\psi_j\rangle\langle\psi_j|. \qquad (5.25)$$

Since $\Pi_j \geq 0$ and together they span the Hilbert space of the qubit, $\sum_{j=0}^2\Pi_j = 1$, we have a legitimate POVM. If we use this POVM and if we get result j, we guess that we were given $|\psi_j\rangle$. The probability of being correct is $p_j = \langle\psi_j|A^\dagger_j A_j|\psi_j\rangle = 2/3$ and the probability of making an error is $q_j = \langle\psi_{j'}|A^\dagger_j A_j|\psi_{j'}\rangle = 1/6$ (for $j \neq j'$). In fact, the above POVM is the optimal one for minimum-error discrimination, p_j takes its maximum possible value, and q_j is the minimum allowed by the laws of quantum mechanics.

For a physical implementation of this optimal POVM along the lines of Neumark's theorem let us define the (non-normalized) qutrit vectors

$$|v_0\rangle = \sqrt{\frac{2}{3}}|0\rangle + \frac{1}{2\sqrt{6}}(|1\rangle + |2\rangle),$$

$$|v_1\rangle = \frac{1}{2\sqrt{2}}(|1\rangle - |2\rangle),$$

$$|u_0\rangle = \frac{1}{2\sqrt{2}}(|1\rangle - |2\rangle),$$

$$|u_1\rangle = \frac{1}{2}\sqrt{\frac{3}{2}}(|1\rangle + |2\rangle), \tag{5.26}$$

where $\{|m_B\rangle\}$ (for $m_B = 0, 1, 2$) is an orthonormal basis in the qutrit Hilbert space which is our $\mathcal{H}_B$. Note that $\langle v_0|v_1\rangle = \langle u_0|u_1\rangle = 0$ and $\|v_0\|^2 + \|v_1\|^2 = \|u_0\|^2 + \|u_1\|^2 = 1$. Let us further introduce the transformation U with the definition,

$$U|0_A\rangle|0_B\rangle = |0_A\rangle|v_{0,B}\rangle + |1_A\rangle|v_{1,B}\rangle,$$

$$U|1_A\rangle|0_B\rangle = |0_A\rangle|u_{0,B}\rangle + |1_A\rangle|u_{1,B}\rangle, \tag{5.27}$$

where A refers to the system and B to the ancilla. As discussed before, it is sufficient to define this transformation on a single initial state $|\psi_B\rangle$ of the ancilla which we conveniently choose as $|\psi_B\rangle = |0_B\rangle$. Then U can be extended to a full unitary transformation on the ancilla space by choosing it to be the identity on the subspace orthogonal to $|0_B\rangle$.

Obviously, any system state can be represented as $|\psi_A\rangle = \alpha|0_A\rangle + \beta|1_A\rangle$ with $|\alpha|^2 + |\beta|^2 = 1$. Multiplying the first of the equations in (5.27) by α and the second by β, adding them together and taking the $|m_B\rangle$ component of the resulting expression, yields

$$\langle m_B|(U(|\psi_A\rangle|0_B\rangle)) = |0_A\rangle(\alpha\langle m|v_0\rangle + \beta\langle m|u_0\rangle) + |1_A\rangle(\alpha\langle m|v_1\rangle + \beta\langle m|u_1\rangle). \tag{5.28}$$

Using $\alpha = \langle 0_A|\psi_A\rangle$ and $\beta = \langle 1_A|\psi_A\rangle$, this expression defines A_m as the operator acting on $|\psi_A\rangle$. Then A_m is explicitly given by

$$A_m = |0_A\rangle(\langle 0_A|\langle m|v_0\rangle + \langle 1_A|\langle m|u_0\rangle + |1_A\rangle(\langle 0_A|\langle m|v_1\rangle + \langle 1_A|\langle m|u_1\rangle). \tag{5.29}$$

Finally, a comparison to Eq. (5.24) reveals that A_j can be written in the compact form,

$$A_j = \sqrt{\frac{2}{3}}|\psi_j\rangle\langle\psi_j|. \tag{5.30}$$

A direct substitution shows that Eq. (5.25) is satisfied by this A_j. Thus, we have just found a physical implementation of the optimal POVM for the minimum-error discrimination of the trine states.

What we have done here can be summarized in the following way. We needed three outcomes of our generalized measurement, so we introduced a three-dimensional ancilla space, called a qutrit. Then we unitarily entangled our system with the ancilla, and, after this interaction, a projective measurement on the ancilla degrees of freedom was performed. The POVM then emerges as the residual effect on the original system due to entanglement of the system to the ancilla when a von Neumann measurement is performed on the ancilla only.

This method corresponds to the tensor product extension of the Hilbert space. The Hilbert space of the combined system is the tensor product of the Hilbert spaces of the two subsystems. There are two conceptually different ways of extending a Hilbert space, the tensor product extension being one of them. The other method is the direct sum extension. The extended Hilbert space is then the direct sum of the Hilbert space spanned by the states of the original system and of the Hilbert space spanned by the auxiliary states, being also called ancilla states. For the trine ensemble, it is possible to associate the three two-dimensional non-normalized detection states $\sqrt{2/3}|\psi_j\rangle$ with three orthonormal states in three dimensions, given by

$$|\tilde{\psi}_j\rangle = \sqrt{\frac{2}{3}}\,|\psi_j\rangle + \sqrt{\frac{1}{3}}\,|2\rangle, \tag{5.31}$$

where the ancillary basis state $|2\rangle$ is orthogonal to the two basis states, $|0\rangle$ and $|1\rangle$, of the system. By performing the von Neumann measurement that consists of the three projections $|\tilde{\psi}_j\rangle\langle\tilde{\psi}_j|$ ($j = 0, 1, 2$) in the enlarged, i.e., three-dimensional Hilbert space, the required generalized measurement is realized in the original two-dimensional Hilbert space of the qubit. In effect, the direct sum extension of the Hilbert space relies on the assumption that the original qubit secretly consists of two components of a qutrit.

5.5 Examples: Strategies for State Discrimination

As examples of a measurement optimization task we will consider two schemes for the optimal discrimination of quantum states. The first is unambiguous discrimination and the second is discrimination with minimum error. We will see that the optimum measurement for the first strategy is a POVM while the optimum measurement for the second is a standard von Neumann measurement. The two main discrimination strategies evolved rather differently from the very beginning. Unambiguous discrimination started with pure states and only very recently was it extended to discriminating among mixed quantum states. Minimum-error discrimination addressed the problem of discriminating between two mixed quantum states from the very beginning and the result for two pure states follows as a special case. The two strategies are, in a sense, complementary to each other. Unambiguous discrimination is relatively straightforward to generalize for more than two states, at least in principle, but it is difficult to treat mixed states. The error-minimizing approach, initially developed for two mixed states, is hard to generalize for more than two states.

5.5.1 Unambiguous Discrimination of Two Pure States

Unambiguous discrimination is concerned with the following problem. An ensemble of quantum systems is prepared so that each individual system is prepared in one of two known states, $|\psi_1\rangle$ or $|\psi_2\rangle$ with probability η_1 or η_2 (such that $\eta_1 + \eta_2 = 1$), respectively. The preparation probabilities are called a priori probabilities or, simply, priors. The states are, in general, not orthogonal, $\langle\psi_1|\psi_2\rangle \neq 0$ but linearly independent. The preparer, Alice, then draws a system at random from this ensemble and hands it over to an observer, called Bob, whose task is to determine which one of the two states he is given. The observer also knows how the ensemble was prepared, i.e., has full knowledge of the two possible states and their priors but does not know the actual state that was drawn. All he can do is to perform a single measurement or perhaps a POVM on the individual system he receives.

In the unambiguous discrimination strategy the observer is not allowed to make an error, i.e., he is not permitted to conclude that he was given one state when actually he was given the other. First we show that this cannot be done with 100% probability of success. To this end, let us assume the contrary and assume we have two detection operators, Π_1 and Π_2, that together span the Hilbert space of the two states,

$$\Pi_1 + \Pi_2 = I. \tag{5.32}$$

For unambiguous detection we also require that

$$\Pi_1|\psi_2\rangle = 0,$$
$$\Pi_2|\psi_1\rangle = 0, \tag{5.33}$$

so that the first detector never clicks for the second state and vice versa, and we can identify the detector clicks with one of the states unambiguously. The probability of successfully identifying the first state if it is given is $p_1 = \langle\psi_1|\Pi_1|\psi_1\rangle$ and the probability of successfully identifying the second state if it is given is $p_2 = \langle\psi_2|\Pi_2|\psi_2\rangle$. Multiplying Eq. (5.32) with $\langle\psi_1|$ from the left and $|\psi_1\rangle$ from the right and taking into account (5.33) give $p_1 = 1$, and, similarly, we obtain $p_2 = 1$, and it appears as though we could have perfect unambiguous discrimination. However, multiplying Eq. (5.32) with $\langle\psi_1|$ from the left and $|\psi_2\rangle$ from the right and taking into account (5.33) again gives $0 = \langle\psi_1|\psi_2\rangle$ which can be satisfied for orthogonal states only. In fact, we have just proved that perfect discrimination of nonorthogonal quantum states is not possible.

Equation (5.32) allows two alternatives only; it assumes that we can have two operators that unambiguously identify the two states all the time. Since this is impossible, we are forced to modify this equation and have to allow for one other alternative. We introduce a third POVM element, Π_0, such that Eq. (5.33) is still satisfied but Eq. (5.32) is modified to

$$\Pi_1 + \Pi_2 + \Pi_0 = I. \tag{5.34}$$

The first and second POVM elements will continue to unambiguously identify the first and second state, respectively. However, Π_0 can click for both states, and, thus, this POVM element corresponds to an inconclusive detection result. It should be emphasized that this outcome is not an error; we will never identify the first state with the second and vice versa; we simply will not make any conclusion in this case. We can now introduce success and failure probabilities in such a way that $\langle \psi_1 | \Pi_1 | \psi_1 \rangle = p_1$ is the probability of successfully identifying $|\psi_1\rangle$ and $\langle \psi_1 | \Pi_0 | \psi_1 \rangle = q_1$ is the probability of failing to identify $|\psi_1\rangle$ (and similarly for $|\psi_2\rangle$). For unambiguous discrimination we have $\langle \psi_2 | \Pi_1 | \psi_2 \rangle = \langle \psi_1 | \Pi_2 | \psi_1 \rangle = 0$ from Eq. (5.33). Using this, we obtain from Eq. (5.34) $p_1 + q_1 = p_2 + q_2 = 1$. This means that if we allow inconclusive detection results to occur with a certain probability then in the remaining cases the observer can conclusively determine the state of the individual system.

It is rather easy to see that a simple von Neumann measurement can accomplish this task. Let us denote the Hilbert space of the two given states by $\mathcal{H}$ and introduce the projector P_1 for $|\psi_1\rangle$ and $\bar{P}_1$ for the orthogonal subspace, such that $P_1 + \bar{P}_1 = I$, the identity in $\mathcal{H}$. Then we know for sure that $|\psi_2\rangle$ was prepared if in the measurement of $\{P_1, \bar{P}_1\}$ a click in the $\bar{P}_1$ detector occurs. A similar conclusion for $|\psi_1\rangle$ can be reached with the roles of $|\psi_1\rangle$ and $|\psi_2\rangle$ reversed. Of course, when a click along P_1 (or P_2) occurs, then we learn nothing about which state was prepared, this outcome thus corresponding to the inconclusive result. In the von Neumann setups one of the alternatives is missing. We either identify one state or we get an inconclusive result, but we miss the other state completely. This scenario is actually allowed by Eq. (5.34).

We now turn our attention to the determination of the optimum measurement strategy for unambiguous discrimination. It is the strategy, or measurement setup, for which the average failure probability is a minimum (or, equivalently, the average success probability is a maximum). We want to determine the operators in Eq. (5.34) explicitly. If we introduce $|\psi_j^\perp\rangle$ as the vector orthogonal to $|\psi_j\rangle$ $(j = 1, 2)$ then the condition of unambiguous detection, Eq. (5.33), mandates the choices

$$\Pi_1 = c_1 |\psi_2^\perp\rangle\langle\psi_2^\perp|, \tag{5.35}$$

and

$$\Pi_2 = c_2 |\psi_1^\perp\rangle\langle\psi_1^\perp|. \tag{5.36}$$

Here c_1 and c_2 are positive coefficients to be determined from the condition of optimality.

Inserting these expressions in the definition of p_1 and p_2 gives $c_1 = p_1 / |\langle\psi_1|\psi_2^\perp\rangle|^2$ and a similar expression for c_2. Finally, introducing $\cos\Theta = |\langle\psi_1|\psi_2\rangle|$ and $\sin\Theta = |\langle\psi_1|\psi_2^\perp\rangle|$, we can write the detection operators as

$$\Pi_1 = \frac{p_1}{\sin^2\Theta} |\psi_2^\perp\rangle\langle\psi_2^\perp|,$$

$$\Pi_2 = \frac{p_2}{\sin^2\Theta} |\psi_1^\perp\rangle\langle\psi_1^\perp|. \tag{5.37}$$

Now, Π_1 and Π_2 are positive semidefinite operators by construction. However, there is one additional condition for the existence of the POVM which is the positivity of the inconclusive detection operator,

$$\Pi_0 = I - \Pi_1 - \Pi_2. \tag{5.38}$$

This is a simple 2×2 matrix in $\mathcal{H}$ and the corresponding eigenvalue problem can be solved analytically. Nonnegativity of the eigenvalues leads, after some tedious but straightforward algebra, to the condition

$$q_1 q_2 \geq |\langle \psi_1 | \psi_2 \rangle|^2, \tag{5.39}$$

where $q_1 = 1 - p_1$ and $q_2 = 1 - p_2$ are the failure probabilities for the corresponding input states.

Equation (5.39) represents the constraint imposed by the positivity requirement on the optimum detection operators. The task we set out to solve can now be formulated as follows. Let

$$Q = \eta_1 q_1 + \eta_2 q_2 \tag{5.40}$$

denote the average failure probability for unambiguous discrimination. We want to minimize this failure probability subject to the constraint, Eq. (5.39). Due to the relation, $P = \eta_1 p_1 + \eta_2 p_2 = 1 - Q$, the minimum of Q also gives us the maximum probability of success. Clearly, for the optimum the product $q_1 q_2$ should be at its minimum allowed by Eq. (5.39), and we can then express q_2 with the help of q_1 as $q_2 = \cos^2 \Theta / q_1$. Inserting this expression in (5.40) yields

$$Q = \eta_1 q_1 + \eta_2 \frac{\cos^2 \Theta}{q_1}, \tag{5.41}$$

where q_1 can now be regarded as the independent parameter of the problem. Optimization of Q with respect to q_1 gives $q_1^{\text{POVM}} = \sqrt{\eta_2 / \eta_1} \cos \Theta$ and $q_2^{\text{POVM}} = \sqrt{\eta_1 / \eta_2} \cos \Theta$. Finally, substituting these optimal values into Eq. (5.40) gives the optimum failure probability,

$$Q^{\text{POVM}} = 2\sqrt{\eta_1 \eta_2} \cos \Theta. \tag{5.42}$$

Let us next see how this result compares to the average failure probabilities of the two possible unambiguously discriminating von Neumann measurements that were described at the beginning of this section. The average failure probability for the first von Neumann measurement, with its failure direction along $|\psi_1\rangle$, can be written by simple inspection as

$$Q_1 = \eta_1 + \eta_2 |\langle \psi_1 | \psi_2 \rangle|^2, \tag{5.43}$$

since $|\psi_1\rangle$ gives a click with probability 1 in this direction, but it is only prepared with probability η_1 and $|\psi_2\rangle$ gives a click with probability $|\langle\psi_1|\psi_2\rangle|^2$, but it is only prepared with probability η_2.

By entirely similar reasoning, the average failure probability for the second von Neumann measurement, with its failure direction along $|\psi_2\rangle$, is given by

$$Q_2 = \eta_1|\langle\psi_1|\psi_2\rangle|^2 + \eta_2. \qquad (5.44)$$

What we can observe is that Q_1 and Q_2 are given as the arithmetic mean of two terms and Q^{POVM} is the geometric mean of the same two terms for either case. So, one would be tempted to say that the POVM performs better always. This, however, is not quite the case; it does so only when it exists. The obvious condition for the POVM solution to exist is that both $q_1^{\mathrm{POVM}} \leq 1$ and $q_2^{\mathrm{POVM}} \leq 1$. Using $\eta_2 = 1 - \eta_1$, a little algebra tells us that the POVM exists in the range $\cos^2\Theta/(1+\cos^2\Theta) \leq \eta_1 \leq 1/(1+\cos^2\Theta)$. If η_1 is smaller than the lower boundary, the POVM goes over to the first von Neumann measurement and if η_1 exceeds the upper boundary the POVM goes over to the second von Neumann measurement. This can be easily seen from Eqs. (5.37) and (5.38) since $p_1 = 1 - q_1 = 0$ for $q_1 = 1$ and Π_0 becomes a projection along $|\psi_1\rangle$ (and correspondingly for $p_2 = 0$).

These findings can be summarized as follows. The optimal failure probability, Q^{opt}, is given as

$$Q^{\mathrm{opt}} = \begin{cases} Q^{\mathrm{POVM}} & \text{if } \frac{\cos^2\Theta}{1+\cos^2\Theta} \leq \eta_1 \leq \frac{1}{1+\cos^2\Theta}, \\ Q_1 & \text{if } \eta_1 < \frac{\cos^2\Theta}{1+\cos^2\Theta}, \\ Q_2 & \text{if } \frac{1}{1+\cos^2\Theta} < \eta_1. \end{cases} \qquad (5.45)$$

The optimum POVM operators are given by

$$\Pi_1 = \frac{1 - q_1^{\mathrm{opt}}}{\sin^2\Theta}|\psi_2^\perp\rangle\langle\psi_2^\perp|,$$

$$\Pi_2 = \frac{1 - q_2^{\mathrm{opt}}}{\sin^2\Theta}|\psi_1^\perp\rangle\langle\psi_2^\perp|. \qquad (5.46)$$

These expressions show explicitly that $\Pi_1 = 0$ and Π_2 is the projector $|\psi_1^\perp\rangle\langle\psi_1^\perp|$ when $q_1^{\mathrm{opt}} = 1$ and $q_2^{\mathrm{opt}} = \cos^2\Theta$, i.e., the POVM goes over smoothly into a projective measurement at the lower boundary and, similarly, into the other von Neumann projective measurement at the upper boundary.

Figure 5.1 displays the failure probabilities, Q_1, Q_2, and Q^{POVM} vs. η_1 for a fixed value of the overlap, $\cos^2\Theta$.

The above result is very satisfying from a physical point of view. The POVM delivers a lower failure probability in its entire range of existence than either of the two von Neumann measurements. At the boundaries of this range it merges smoothly with the one von Neumann measurement that has a lower failure

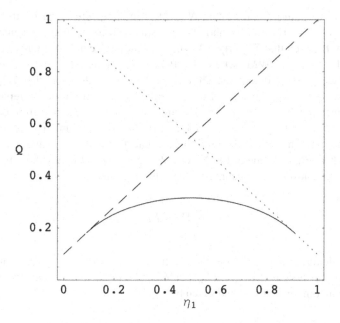

Fig. 5.1 Failure probability, Q, vs. the prior probability, η_1. *Dashed line*: Q_1, *dotted line*: Q_2, *solid line*: Q^{POVM}. For the figure we used the following representative value: $|\langle \psi_1 | \psi_2 \rangle|^2 = 0.1$. For this the optimal failure probability, Q_{opt} is given by Q_1 for $0 < \eta_1 < 0.09$, by Q^{POVM} for $0.09 \leq \eta_1 \leq 0.91$, and by Q_2 for $0.91 < \eta_1$

probability at that point. Outside this range the state preparation is dominated by one of the states and the optimal measurement becomes a von Neumann projective measurement, using the state that is prepared less frequently as its failure direction.

5.5.2 Minimum-Error Discrimination of Two Quantum States

In the previous section we have required that, whenever a definite answer is returned after a measurement on the system, the result should be unambiguous, at the expense of allowing inconclusive outcomes to occur. For many applications in quantum communication, however, one wants to have conclusive results only. This means that errors are unavoidable when the states are nonorthogonal. Based on the outcome of the measurement, in each single case then a guess has to be made as to what the state of the quantum system was. In the optimal strategy we want to minimize the probability of making a wrong guess; hence, this procedure is known as *minimum-error* discrimination. The problem is to find the optimum measurement that minimizes the probability of errors.

Let us state the optimization problem a little more precisely. In the most general case, we want to distinguish, with minimum probability of error, among N given states of a quantum system (where $N \geq 2$). The states are given by the density operators ρ_j ($j = 1, 2, \ldots, N$) and the jth state occurs with the given a priori probability η_j, such that $\sum_{j=1}^{N} \eta_j = 1$. The measurement can be formally described with the help of a POVM, where the POVM elements, Π_j, correspond to the possible measurement outcomes. They are defined in such a way that $\mathrm{Tr}(\rho \Pi_j)$ is the probability to infer the state of the system to be ρ_j if it has been prepared in a state ρ. Since the probability is a real nonnegative number, the detection operators once again have to be positive semidefinite. In the error-minimizing measurement scheme the measurement is required to be exhaustive and conclusive in the sense that in each single case one of the N possible states is identified with certainty and inconclusive results do not occur. This leads to the requirement

$$\sum_{j=1}^{N} \Pi_j = I_{D_S}, \tag{5.47}$$

where I_{D_S} denotes the identity operator in the D_S-dimensional physical state space of the quantum system. The overall probability P_{err} to make an erroneous guess for any of the incoming states is then given by

$$P_{\mathrm{err}} = 1 - P_{\mathrm{corr}} = 1 - \sum_{j=1}^{N} \eta_j \mathrm{Tr}(\rho_j \Pi_j) \tag{5.48}$$

with $\sum_j \eta_j = 1$. Here we introduced the probability P_{corr} that the guess is correct. In order to find the minimum-error measurement strategy, one has to determine the POVM that minimizes the value of P_{err} under the constraint given by Eq. (5.47). By inserting these optimum detection operators into Eq. (5.48), the minimum-error probability $P_{\mathrm{err}}^{\mathrm{min}} \equiv P_E$ is determined. The explicit solution to the error-minimizing problem is not trivial and analytical expressions have been derived only for a few special cases.

For the case that only two states are given, either pure or mixed, the minimum-error probability, P_E, was derived in the mid-1970s by Helstrom in the framework of quantum detection and estimation theory. We find it more instructive to start by analyzing the two-state minimum-error measurement with the help of an alternative method that allows us to gain immediate insight into the structure of the optimum detection operators, without applying variational techniques. Starting from Eq. (5.48) and making use of the relations $\eta_1 + \eta_2 = 1$ and $\Pi_1 + \Pi_2 = I_{D_S}$ that have to be fulfilled by the a priori probabilities and the detection operators, respectively, we see that the total probability to get an erroneous result in the measurement is given by

$$P_{\mathrm{err}} = 1 - \sum_{j=1}^{2} \eta_j \mathrm{Tr}(\rho_j \Pi_j) = \eta_1 \mathrm{Tr}(\rho_1 \Pi_2) + \eta_2 \mathrm{Tr}(\rho_2 \Pi_1). \tag{5.49}$$

This can be alternatively expressed as

$$P_{err} = \eta_1 + \text{Tr}(\Lambda\Pi_1) = \eta_2 - \text{Tr}(\Lambda\Pi_2), \qquad (5.50)$$

where we introduced the hermitian operator

$$\Lambda = \eta_2\rho_2 - \eta_1\rho_1 = \sum_{k=1}^{D_S} \lambda_k |\phi_k\rangle\langle\phi_k|. \qquad (5.51)$$

Here the states $|\phi_k\rangle$ denote the orthonormal eigenstates belonging to the eigenvalues λ_k of the operator Λ. The eigenvalues are real, and without loss of generality we can number them in such a way that

$$\lambda_k < 0 \qquad \text{for} \qquad 1 \le k < k_0,$$
$$\lambda_k > 0 \qquad \text{for} \qquad k_0 \le k \le D,$$
$$\lambda_k = 0 \qquad \text{for} \qquad D < k \le D_S. \qquad (5.52)$$

By using the spectral decomposition of Λ, we get the representations

$$P_{err} = \eta_1 + \sum_{k=1}^{D_S} \lambda_k\langle\phi_k|\Pi_1|\phi_k\rangle = \eta_2 - \sum_{k=1}^{D_S} \lambda_k\langle\phi_k|\Pi_2|\phi_k\rangle. \qquad (5.53)$$

Our optimization task now consists in determining the specific operators Π_1, or Π_2, respectively, that minimize the right-hand side of Eq. (5.53) under the constraint that

$$0 \le \langle\phi_k|\Pi_j|\phi_k\rangle \le 1 \qquad (j = 1,2) \qquad (5.54)$$

for all eigenstates $|\phi_k\rangle$. The latter requirement is due to the fact that $\text{Tr}(\rho\Pi_j)$ denotes a probability for any ρ. From this constraint and from Eq. (5.53) it immediately follows that the smallest possible error probability, $P_{err}^{min} \equiv P_E$, is achieved when the detection operators are chosen in such a way that the equations $\langle\phi_k|\Pi_1|\phi_k\rangle = 1$ and $\langle\phi_k|\Pi_2|\phi_k\rangle = 0$ are fulfilled for eigenstates belonging to negative eigenvalues, while eigenstates corresponding to positive eigenvalues obey the equations $\langle\phi_k|\Pi_1|\phi_k\rangle = 0$ and $\langle\phi_k|\Pi_2|\phi_k\rangle = 1$. Hence the optimum POVM operators can be written as

$$\Pi_1 = \sum_{k-1}^{k_0-1} |\phi_k\rangle\langle\phi_k|, \qquad \Pi_2 = \sum_{k=k_0}^{D_S} |\phi_k\rangle\langle\phi_k|, \qquad (5.55)$$

where the expression for Π_2 has been supplemented by projection operators onto eigenstates belonging to the eigenvalue $\lambda_k = 0$, in such a way that $\Pi_1 + \Pi_2 = I_{D_S}$. Obviously, provided that there are positive as well as negative eigenvalues in the spectral decomposition of Λ, the minimum-error measurement for discriminating two quantum states is a von Neumann measurement that consists in performing projections onto the two orthogonal subspaces spanned by the set of states $\{|\phi_1\rangle, \ldots, |\phi_{k_0-1}\rangle\}$, on the one hand, and $\{|\phi_{k_0}\rangle, \ldots, |\phi_{D_S}\rangle\}$, on the other hand. An interesting special case arises when negative eigenvalues do not exist. In this case it follows that $\Pi_1 = 0$ and $\Pi_2 = I_{D_S}$ which means that the minimum error probability can be achieved by always guessing the quantum system to be in the state ρ_2, without performing any measurement at all. Similar considerations hold true in the absence of positive eigenvalues, so a measurement does not always aid minimum-error discrimination. By inserting the optimum detection operators into Eq. (5.50), the minimum-error probability is found to be

$$
P_E = \eta_1 - \sum_{k=1}^{k_0-1} |\lambda_k| = \eta_2 - \sum_{k=k_0}^{D} |\lambda_k|.
\tag{5.56}
$$

Taking the sum of these two alternative representations and using $\eta_1 + \eta_2 = 1$, we arrive at

$$
P_E = \frac{1}{2}\left(1 - \sum_k |\lambda_k|\right) = \frac{1}{2}(1 - \mathrm{Tr}|\Lambda|),
\tag{5.57}
$$

where $|\Lambda| = \sqrt{\Lambda^\dagger \Lambda}$. Together with Eq. (5.48) this immediately yields the well-known Helstrom formula for the minimum-error probability in discriminating ρ_1 and ρ_2,

$$
P_E = \frac{1}{2}(1 - \mathrm{Tr}|\eta_2\rho_2 - \eta_1\rho_1|) = \frac{1}{2}(1 - \|\eta_2\rho_2 - \eta_1\rho_1\|).
\tag{5.58}
$$

In the special case that the states to be distinguished are the pure states $|\psi_1\rangle$ and $|\psi_2\rangle$, this expression reduces to

$$
P_E = \frac{1}{2}\left(1 - \sqrt{1 - 4\eta_1\eta_2 |\langle\psi_1|\psi_2\rangle|^2}\right).
\tag{5.59}
$$

This expression, which is the one found in textbooks, can be cast to the equivalent form,

$$
P_E = \eta_{\min}\left(1 - \frac{2\eta_{\max}(1 - |\langle\psi_1|\psi_2\rangle|^2)}{\eta_{\max} - \eta_{\min} + \sqrt{1 - 4\eta_{\min}\eta_{\max}|\langle\psi_1|\psi_2\rangle|^2}}\right),
\tag{5.60}
$$

where $\eta_{\min}$ ($\eta_{\max}$) is the smaller (greater) of the prior probabilities, η_1 and η_2. This form lends itself to a transparent interpretation. The first factor on the right-hand side is what we would get if we always guessed the state that is prepared more often, without any measurement at all. Thus, the factor multiplying $\eta_{\min}$ is the result of the optimized measurement.

The setup of the detectors that achieve the optimum error probabilities is particularly simple for the case of equal a priori probabilities. Two orthogonal detectors, placed symmetrically around the two pure states, will do the task. The simplicity is particularly striking when one compares this setup to the corresponding POVM setup for optimal unambiguous discrimination.

Finally, we present an interesting relation, without proof, that is always satisfied between the minimum-error probability of the minimum-error detection and the optimal failure probability of unambiguous detection. It reads as

$$P_E \le \frac{1}{2} Q^{\text{opt}}. \tag{5.61}$$

This means that for two arbitrary states (mixed or pure), prepared with arbitrary a priori probabilities, the smallest possible failure probability in unambiguous discrimination is at least twice as large as the smallest probability of errors in minimum-error discrimination of the same states.

5.6 Problems

1. Find the eigenvalues of the POVM element, given in Eq. (5.38), corresponding to inconclusive outcomes and show that the condition of their positivity can be cast to the form given in Eq. (5.39).
2. For optimum unambiguous discrimination between two pure quantum states, the POVM elements are given in Eq. (5.46) and by $\Pi_0 = I - \Pi_1 - \Pi_2$. Find an implementation in terms of a generalized measurement via Neumark's theorem, introducing the ancilla by using the tensor product extension of the Hilbert space.
3. The derivation of the general formula for the minimum-error probability is given in the text.

   (a) Show that for the special case of two pure states, $\rho_1 = |\psi_1\rangle\langle\psi_1|$ and $\rho_2 = |\psi_2\rangle\langle\psi_2|$, Eq. (5.58) reduces to Eq. (5.59).
   (b) The general expression for the optimal detection operators is given in Eq. (5.55). Find their explicit expression for the pure state case of part (a).

4. (a) Show that Q^{POVM} in Eq. (5.42) and P_E in Eq. (5.59) satisfy the inequality (5.61).
   (b) If you are very ambitious, prove the inequality.
5. (a) Let us consider the so-called trine states

$$|\psi_1\rangle = |0\rangle \quad |\psi_2\rangle = -\tfrac{1}{2}(|0\rangle + \sqrt{3}|1\rangle)$$
$$|\psi_3\rangle = -\tfrac{1}{2}(|0\rangle - \sqrt{3}|1\rangle).$$

These are states of a single qubit. We are given a qubit that is guaranteed to be in one of these three states, and we want to find a POVM that does the

following. If we obtain result 1 (corresponding to operators A_1 and $A_1^\dagger$), then
we know that the qubit we were given was not in state $|\psi_1\rangle$. If we get result 2,
it was not in state $|\psi_2\rangle$, and if we get result 3, then it was not in state $|\psi_3\rangle$.
Find a POVM that does this.

(b) Now let us look at the four states (the tetrad states)

$$|\psi_1\rangle = \frac{1}{\sqrt{3}}(-|0\rangle + \sqrt{2}e^{-2\pi i/3}|1\rangle) \ |\psi_2\rangle = \frac{1}{\sqrt{3}}(-|0\rangle + \sqrt{2}e^{2\pi i/3}|1\rangle)$$

$$|\psi_3\rangle = \frac{1}{\sqrt{3}}(-|0\rangle + \sqrt{2}|1\rangle) \qquad\qquad |\psi_4\rangle = |0\rangle.$$

We want to consider the minimum-error detection scenario for these states.
That is, we are given a qubit in one of these four states, and we want to find
which one, with the requirement that our probability of making a mistake
is the smallest possible. The POVM that accomplishes this is given by the
operators $A_j = (1/\sqrt{2})|\psi_j\rangle\langle\psi_j|$, where $j = 1,\ldots,4$. Verify that

$$\sum_{j=1}^{4} A_j^\dagger A_j = I,$$

and find the probability that a state will be correctly identified. Also find the
probability that an error will be made, that is, that we are given $|\psi_j\rangle$ but we
identify it as $|\psi_{j'}\rangle$, where $j \neq j'$.

6. When we derived POVMs we used a Hilbert space that was a tensor product
between the space for the system we wanted to measure and the space for an
ancilla. It is also possible to derive a POVM by considering the direct sum of
two Hilbert spaces. In particular, if we are measuring states that are confined to a
subspace of a larger space, then we can describe projective measurements on the
entire space as POVMs on the subspace. Let us see how this works by means of
an example.

(a) Consider again the trine states, but now let us suppose that they are states
of a qutrit rather than of a qubit. The entire Hilbert space, $\mathcal{H}_3$, has the
orthonormal basis, $\{|0\rangle, |1\rangle, |2\rangle\}$, and the subspace, S, in which the trine
states lie, is spanned by the basis elements $|0\rangle$ and $|1\rangle$. The POVM operators
for the minimum-error scenario are given by $A_j = \sqrt{2/3}|\psi_j\rangle\langle\psi_j|$ (the states
$|\psi_j\rangle$, for $j = 1, 2, 3$ are given in part (a) of problem 1. Find one-dimensional
projections P_j, acting on $\mathcal{H}_3$ that satisfy

$$\langle\psi|P_j|\psi\rangle = \langle\psi|A_j^\dagger A_j|\psi\rangle,$$

for any state $|\psi\rangle \in S$.

(b) Suppose that we want to measure the projections P_j and that we can easily measure the projections corresponding to the basis states $\{|0\rangle, |1\rangle, |2\rangle\}$. We can then measure the projections P_j by measuring the projections $|j\rangle\langle j|$ if we can find a unitary transformation, U, such that $|j\rangle\langle j| = UP_jU^{-1}$. This implies that

$$|\langle j|U|\psi\rangle|^2 = \langle\psi|P_j|\psi\rangle,$$

so that the probability of measuring $|j\rangle$ in the transformed state is the same as that of measuring P_j in the original state. Find such a unitary operator in this case.

References

1. J. Preskill, *Lecture Notes for Physics 219*, http://www.theory.caltech.edu/people/preskill/ph229/
2. J.A. Bergou, Tutorial review: Discrimination of quantum states. J. Mod. Opt. **57**, 160 (2010)
3. S.M. Barnett, S. Croke, Quantum state discrimination. Adv. Opt. Photo. **1**, 238 (2009)

Chapter 6
Quantum Cryptography

6.1 Outline

Quantum communication is the most advanced area of quantum information processing and quantum computing. This is where the most fundamental features of quantum mechanics are only a short step away from spectacular practical applications. We have already seen two such applications: dense coding and teleportation. In this chapter we shall deal with what is arguably the most successful area of all of quantum information and quantum computing: quantum cryptography.

Cryptography is the art of secret communication. It has been around since ancient times. What distinguishes quantum cryptography from classical cryptography is that classical information can be copied at will. Therefore, no classical cryptographic protocol is entirely secure, although there are classical cryptographic protocols that are very hard to break in practice. Quantum information, on the other hand, i.e., unknown quantum states, cannot be cloned (cf. the no-cloning theorem). Quantum cryptography enables two parties, traditionally called Alice and Bob, to exchange information in a provably secure way. The security stems from the fact that if an eavesdropper, traditionally called Eve, tries to intercept the messages, her presence can be detected by the unavoidable disturbance she causes by trying to access the information.

In the next section of this chapter we give a very brief introduction to the ideas behind classical cryptography. At the heart of any provably secure cryptographic protocol lies the process of establishing a secret key. Quantum key distribution (QKD) solves this problem using fundamental principles of quantum mechanics. Alice and Bob can then use the secret key to encode and decode their messages. In the following two sections we will briefly describe a number of QKD protocols that clearly show what fundamental features of quantum mechanics are being used as resources.

Quantum cryptography is now a highly developed subject, and what we will present here just scratches the surface. Our intent is to provide an introduction to some of the basic ideas on which the subject is based.

J.A. Bergou and M. Hillery, *Introduction to the Theory of Quantum Information Processing*, Graduate Texts in Physics, DOI 10.1007/978-1-4614-7092-2_6,

6.2   The One-Time Pad

The first documented cases of secret communication date back about thirty centuries. Since then its history can be described as the ongoing struggle between code makers and code breakers. Sometimes, code makers outsmart the code breakers; sometimes the code breakers are ahead in the game. With quantum mechanics, code makers finally seem to be gaining the upper hand.

To be precise, the word code refers to the particular kind of secret communication where a word or even a full sentence is replaced with a word, a number, or a symbol. Initially very popular, its use has decreased over time to give way to the cipher, which acts at the level of the smallest building blocks: the letters. In a cipher letters are being replaced by letters, numbers, or symbols.

If letters of a message are simply rearranged, we are talking about transposition. The alternative to transposition is substitution when letters are being substituted. An early example for a substitution cipher is the *Caesar shift*, used by Julius Caesar for the purpose of his military correspondence. In this cipher each letter in the message is simply replaced with the letter that is three places further down the alphabet. So, for example, *Caesar* becomes *Fdhvdu*.

This code is very easy to break. We can make things a little more difficult for an eavesdropper if we use different shifts for different messages, that is, we do not always shift by three. Then, in addition to the message, the shift has to be sent to the receiver. This is the simplest example of a key. The key, in this case the shift, enables the receiver to decrypt the message. Things get much harder for the eavesdropper if we use different shifts for different letters in a single message. This, of course, makes the key much longer; rather than one number (shift) for each message, we have one for each letter in a message. The advantage, however, is that if we have a random key, and only use it once, the code is unbreakable. This procedure is known as a one-time pad.

This simple example for a cipher shows two distinct ingredients of encryption: the algorithm and the key. The algorithm specifies the encryption and decryption procedures, but in order to use it, one has to have a key known by both the sender and receiver. The algorithm can be publicly known, so that the security of the system depends on restricting knowledge of the key. Thus, a major problem in cryptography is how to distribute a secure key to only the legitimate users. This is the problem of key distribution. One can resort to couriers with briefcases handcuffed to their wrists, but in the electronic age, something more sophisticated is needed. What we will show in this chapter is that quantum mechanics can be used to distribute secure keys. The resulting prescriptions are known as QKD protocols.

6.3   The B92 Quantum Key Distribution Protocol

Both of the state discrimination strategies discussed in the previous chapter come very nicely together in the so-called B92 QKD protocol. In 1992 Charles Bennett proposed using the unambiguous discrimination of two nonorthogonal states as the basis of a form of quantum cryptography. Quantum cryptography is a method of generating a secure shared key by quantum mechanical means that is discarded after being used only once. So, it is the quantum version of the one-time pad cipher.

In cryptography the sender is often called Alice and the receiver Bob. We will use this nomenclature in what follows. As we have seen, the question is, then, how to generate, or distribute, a secret key between Alice and Bob. The B92 protocol provides one possible solution to this problem, using quantum mechanical means. Here is how it works:

1. Alice generates a random sequence of 0s and 1s, that is, a random classical bit string.
2. Alice encodes each data bit in a qubit, $|\psi_0\rangle = |0\rangle$ if the corresponding bit is 0, and $|\psi_1\rangle = \frac{1}{\sqrt{2}}(|0\rangle + |1\rangle)$ if the corresponding bit is 1. This way she generates a random string of qubits.
3. Alice then sends the resulting string of qubits to the receiver, Bob.
4. Bob applies optimum unambiguous state discrimination strategy to each qubit he receives. Using Eq. (5.42), the success probability for Bob's measurement is $P = 1 - Q^{\text{POVM}} = 1 - \frac{1}{\sqrt{2}} \approx 0.293$.
   From now on, Alice and Bob will exchange only classical information.
5. Bob tells Alice, over a public classical channel, in which instances the discrimination succeeded but not the result.
6. They keep only those bits when the discrimination was successful and delete those when it failed. After this they share the so-called *raw key*.

The raw key is the same for Alice and Bob, since Alice knows what she sent in those instances when Bob successfully identified the state of the qubit and Bob was using unambiguous discrimination, so there is no error. This is true as long as an eavesdropper and noise are both absent.

So, why is this procedure secure if there is an eavesdropper? Suppose the eavesdropper, called Eve, has intercepted a qubit. She cannot determine whether it is in the state $|\psi_0\rangle$ or $|\psi_1\rangle$. One thing she can do is to apply the optimum unambiguous state discrimination procedure. Then she will fail with a probability of $\frac{1}{\sqrt{2}} \approx 71\%$. When she does, she has no idea what state was sent, so she must guess which one to send to Bob. Since the two states are prepared with equal probability, Eve will guess half the time right and half the time wrong. This means that the probability that Bob will receive a wrong bit is $\frac{1}{2\sqrt{2}} \approx 35.3\%$. These errors can easily be detected if Alice and Bob add one more step to their protocol.

7'. Alice and Bob publicly compare some of their bits. If there are no errors there is no eavesdropper and they keep the remaining bits. If there are errors, in the range of 35%, there is likely to be an eavesdropper. They then simply throw out all bits and try again.

But wait a second. Eve's goal, besides learning as much as possible about the key, is also to introduce as few errors as possible. There are unavoidable errors in any communication scheme, partly due to the imperfections of the communication channel and partly due to the imperfect detection. Eve's goal is to remain below this unavoidable noise level in order to avoid being detected. So, suppose she has intercepted the particle but she now chooses the minimum-error strategy to determine which state was sent. Using Eq. (5.59), her error rate will now be $\frac{1}{2}(1 - \frac{1}{\sqrt{2}}) \approx 14.6\%$ which is much less than the error rate that she introduces if she uses the unambiguous discrimination strategy. In addition, she still learns the key with a fidelity of about 85%. However, even this rather low error rate can still be detected if Alice and Bob modify the last step of their protocol.

7. Over a classical communication channel, Alice and Bob publicly compare some of their bits. If there are no errors there is no eavesdropper and they keep the remaining bits. If there are errors, in the range of 14%, there is likely to be an eavesdropper. They then simply throw out all bits and try again.

This requirement is much more stringent than the one in Step 7 of the original protocol. It is still possible to detect the presence of an eavesdropper but the requirements on the channel quality and detector efficiency are much more demanding than in the case when Eve uses the same strategy as Bob. So, here we had an example where one state discrimination strategy is optimal for the intended recipient and the other for the eavesdropper and to analyze the worst case scenario for Alice and Bob we have to consider all of their possibilities. There are many other QKD protocols but this one is perhaps the clearest example of how important optimal detection strategies are for quantum communication.

6.4   The BB 84 Protocol

The first, and most famous, QKD protocol was developed by Bennett and Brassard in 1984, and is known as the BB 84 protocol. It is actually possible to buy commercial quantum cryptography systems that make use of this protocol. It is a four-state protocol in which Alice and Bob make use of two sets of bases to establish a shared key.

Alice sends qubits to Bob, each state being a member of one of two sets of orthonormal bases, the z basis, $\{|0\rangle, |1\rangle\}$ or the x basis $\{|+x\rangle, |-x\rangle\}$, where $|\pm x\rangle = (|0\rangle \pm |1\rangle)/\sqrt{2}$. She decides which state to send at random, i.e. she chooses a basis at random and a state from that basis at random. The states $|0\rangle$ and $|+x\rangle$ correspond to a bit value of 0 and $|1\rangle$ and $|-x\rangle$ correspond to a bit value of 1. Upon receiving the qubit, Bob measures it in one of the two bases, choosing which basis to use at random. If he uses the same basis as the one Alice chose, he will obtain the same state that Alice sent. For example, if Alice sends $|0\rangle$ and Bob measures in the

z basis, he will obtain $|0\rangle$. If, however, Bob chooses the wrong basis, his results will be random. If Alice sent $|0\rangle$ and Bob measures in the x basis, he will obtain $|+x\rangle$ with a probability of $1/2$ and $|-x\rangle$ with a probability of $1/2$. After measuring a qubit, Bob announces over a public channel which basis he used, but not the result of the measurement. Alice then tells Bob whether the basis he used was the same as the one she chose. If they agree, they keep the bit value corresponding to that qubit. If they disagree, they throw out that bit.

In the intercept-resend attack, the eavesdropper, Eve, captures the qubit that Alice sent, measures it, and then, based on her measurement result, prepares another qubit to send on to Bob. Her problem is that she does not know in which basis to measure Alice's qubit, so she has to guess. If she guesses correctly, and Alice and Bob use the same basis, she knows the value of that key bit, and she has not been detected. However, with a probability of $1/2$ she will guess incorrectly, and measure the qubit in the wrong basis, and obtain a random result, which she will then use to prepare a qubit to send to Bob. For example, suppose Alice sends $|0\rangle$ but Eve measures in the x basis. She will, with a probability of $1/2$, obtain $|+x\rangle$ and send that on to Bob, and will obtain $|-x\rangle$, also with a probability of $1/2$, and send that on to Bob. Now suppose that Bob chooses the same basis as Alice, in this case the z basis. In either case, whether Eve sent $|+x\rangle$ or $|-x\rangle$, he will obtain $|0\rangle$ with a probability of $1/2$ and $|1\rangle$ with a probability of $1/2$. If he obtains $|0\rangle$, then Eve's intervention goes undetected, and she knows the value of the bit. However, if Bob obtains $|1\rangle$, when he should have obtained $|0\rangle$ had there been no eavesdropper, Eve's presence will be revealed. Consequently, Eve will introduce errors in the case when Alice and Bob use the same basis. This will happen with a probability of $1/4$; a probability of $1/2$ that Eve chooses the wrong basis times a probability of $1/2$ that if she does, Bob obtains a measurement result different from what Alice sent. Alice and Bob can detect these errors by publicly comparing a subset of the bits for which they chose the same bases. If there are no errors, there was no eavesdropping, but if there are errors, there was an eavesdropper present. In that case they throw out all of the bits and start over.

Eve can try a different kind of attack in which she entangles the incoming qubit with an ancilla. When she receives Alice's qubit, she appends an ancilla qubit to it in the state $|0\rangle$ and then applies a two-qubit unitary operation, U, which acts as

$$U|0\rangle_a|0\rangle_e = |0\rangle_a|\phi_{00}\rangle_e + |1\rangle_a|\phi_{01}\rangle_e$$
$$U|1\rangle_a|0\rangle_e = |0\rangle_a|\phi_{10}\rangle_e + |1\rangle_a|\phi_{11}\rangle_e, \tag{6.1}$$

where the subscript a designates Alice's qubit and the subscript e designates Eve's qubit. Because U is unitary, the states of Eve's qubit must satisfy

$$\|\phi_{00}\|^2 + \|\phi_{01}\|^2 = 1 \qquad \|\phi_{10}\|^2 + \|\phi_{11}\|^2 = 1$$
$$\langle\phi_{00}|\phi_{10}\rangle + \langle\phi_{01}|\phi_{11}\rangle = 0. \tag{6.2}$$

After entangling her qubit with Alice's, Eve sends Alice's qubit on to Bob.

We will not do a complete analysis of this attack, but we will show that if Eve is to introduce no errors, then she can obtain no information. If no errors are to be produced when Bob measures in the z basis, then we must have $|\phi_{01}\rangle = |\phi_{10}\rangle = 0$. Now let us see what happens in the x basis. First, we have that

$$U|\pm x\rangle_a|0\rangle_e = \frac{1}{\sqrt{2}}[|0\rangle_a(|\phi_{00}\rangle_e \pm |\phi_{10}\rangle_e)$$

$$+|1\rangle_a(|\phi_{01}\rangle_e \pm |\phi_{11}\rangle_e)]. \tag{6.3}$$

Now if there are to be no errors when Bob measures in the x basis, then when Alice's qubit is $|+x\rangle_a$, the right-hand side of the above equation must be proportional to $|+x\rangle$, which implies that

$$|\phi_{00}\rangle_e + |\phi_{10}\rangle_e = |\phi_{01}\rangle_e + |\phi_{11}\rangle_e, \tag{6.4}$$

and when Alice sends $|-x\rangle_a$, the state after applying U should be proportional to $|-x\rangle$, which implies that

$$|\phi_{00}\rangle_e - |\phi_{10}\rangle_e = -(|\phi_{01}\rangle_e - |\phi_{11}\rangle_e). \tag{6.5}$$

Combining these conditions with the one we obtained from the z basis, $|\phi_{01}\rangle = |\phi_{10}\rangle = 0$, we see that we must also have $|\phi_{00}\rangle = |\phi_{11}\rangle$. These two conditions, however, imply that Eve's qubit is not entangled with Alice's qubit at all, and, therefore, Eve transfers no information about Alice's qubit to her ancilla qubit. Thus, if Eve is to introduce no errors, she will gain no information.

6.5   The E91 Protocol

In 1991 Artur Ekert proposed a protocol based on shared entanglement rather than on one party sending particles directly to another. Suppose a source sends one qubit to Alice and another to Bob and suppose that these qubits are in a singlet state. Alice and Bob, independently and randomly, decide to whether to measure their qubit in the x or y bases, where $|\pm y\rangle = (|0\rangle \pm i|1\rangle)/\sqrt{2}$. Alice and Bob then announce which basis they used. If they used the same basis, their result will be anticorrelated, e.g., if Alice got $|+x\rangle$, Bob will have gotten $|-x\rangle$. Since each knows what the other got, they can use this information to establish a key.

Ekert also proposed that they use their measurement results for the cases in which they chose different bases to test whether a Bell inequality is violated or not. If Eve had taken over the source and were sending Alice and Bob particles in definite states, for example, a $|+x\rangle$ to Alice and a $|-x\rangle$ to Bob, then the Bell inequality would not be violated, and Alice and Bob would detect her.

6.6 Quantum Secret Sharing

Secret sharing is a cryptographic protocol in which a secret is split into several parts with each part being given to a different party. In order to recover the secret, all of the parties have to cooperate. It is a means of providing extra security. For example, a bank manager may split the combination of the vault into two pieces and give each piece to a different person. The reasoning is that if at least one of the persons is honest, an honest person will keep a dishonest one from doing anything wrong once the vault is open. If both people are dishonest, this will not work, but the probability of encountering two dishonest people is lower than encountering one, so an extra measure of security is gained.

Classically one can split a key very easily. Suppose Alice possesses a sequence of zeroes and ones, which she wants to use as a key. She creates a random sequence of zeroes and ones and adds it, bitwise and modulo 2, to the key sequence, to create, what we will call, a sum sequence. She sends the sum sequence to Bob and the random sequence to Charlie. In order to find Alice's key sequence, Bob and Charlie have to cooperate. In particular, if Bob and Charlie add their sequences bitwise and modulo 2, the random sequence cancels out and they are left with Alice's original sequence.

One can combine this procedure with QKD to form a quantum secret sharing protocol that provides protection against eavesdropping. Alice uses, for example, BB84, to establish keys with Bob and Charlie. The actual key she will use to encode any messages is just the sum (bitwise and modulo 2) of these two keys. Therefore, in order to decode any message that Alice sends them, Bob and Charlie will have to cooperate, in particular they will have to combine their two keys to find the one Alice is actually using.

Another way of approaching this problem is to use entanglement. Suppose Alice prepares one of two entangled states

$$|\Psi_0\rangle = \cos\theta|00\rangle + \sin\theta|11\rangle$$
$$|\Psi_1\rangle = \cos\theta|00\rangle - \sin\theta|11\rangle, \tag{6.6}$$

where $|\Psi_0\rangle$ corresponds to a bit value of 0 and $|\Psi_1\rangle$ corresponds to a bit value of 1. She sends one qubit to Bob and the other to Charlie. As we will see, Bob and Charlie have to cooperate in order to find out which state Alice sent. Bob now measures his qubit in the x basis. Define the single qubit states

$$|\psi_\perp\rangle = \cos\theta|0\rangle \pm \sin\theta|1\rangle. \tag{6.7}$$

If Alice sent $|\Psi_0\rangle$, then if Bob gets $|+x\rangle$, Charlie will have the state $|\psi_+\rangle$, and if Bob gets $|-x\rangle$, then Charlie will have the state $|\psi_-\rangle$. Similarly, if Alice sent $|\Psi_1\rangle$, then if Bob gets $|+x\rangle$, Charlie will have the state $|\psi_-\rangle$, and if Bob gets $|-x\rangle$, then Charlie will have the state $|\psi_+\rangle$. Charlie now performs optimal unambiguous state discrimination for the states $|\psi_\pm\rangle$ on his qubit. He will succeed with a probability of

$1 - |\cos(2\theta)|$. He tells Alice and Bob when his measurement succeeds and when it fails. They throw out the instances in which it failed. In the case in which Charlie's measurement succeeds, he has either the result $|\psi_+\rangle$ or the result $|\psi_-\rangle$, and Bob has either $|+x\rangle$ or $|-x\rangle$. Neither of them alone can determine which state Alice sent, but if they combine their results, they can. Therefore, the information about which state, and thereby which key bit, Alice sent is split between Bob and Charlie.

An eavesdropper is faced with the same situation as in the B92 protocol, distinguishing between two nonorthogonal states, in this case $|\Psi_0\rangle$ and $|\Psi_1\rangle$. Eve will invariably misidentify the state she receives some of the time, and she will then send the wrong state on to Bob and Charlie. That will lead to the situation in which Bob and Charlie receive a different state from the one that Alice sent, and this will result in errors in the shared key. These errors can be detected if Alice, Bob, and Charlie compare a subset of their key bits. If there are no errors, there was no eavesdropper present, and the key is secure.

6.7   Problems

1. Suppose we are using a singlet state $|\phi_-\rangle = (|0\rangle|1\rangle - |1\rangle|0\rangle)/\sqrt{2}$ in the Ekert 91 protocol, and Alice and Bob are measuring in the x and y bases. We want to find a Bell inequality that is maximally violated under these conditions. Suppose the two observables that Alice is using in the Bell inequality are σ_x and σ_y. Find two observables for Bob of the form, $\hat{\mathbf{n}}_1 \cdot \boldsymbol{\sigma}$ and $\hat{\mathbf{n}}_2 \cdot \boldsymbol{\sigma}$, where $\hat{\mathbf{n}}_1$ and $\hat{\mathbf{n}}_2$ are unit vectors in the $x - y$ plane such that the expression appearing in the resulting Bell inequality

$$|\langle \sigma_x(\hat{\mathbf{n}}_1 \cdot \boldsymbol{\sigma})\rangle + \langle \sigma_x(\hat{\mathbf{n}}_2 \cdot \boldsymbol{\sigma})\rangle + \langle \sigma_y(\hat{\mathbf{n}}_1 \cdot \boldsymbol{\sigma})\rangle - \langle \sigma_y(\hat{\mathbf{n}}_2 \cdot \boldsymbol{\sigma})\rangle|$$

is equal to $2\sqrt{2}$ for the singlet state. It is useful to first prove that for any unit vectors $\hat{\mathbf{e}}$ and $\hat{\mathbf{n}}$, $\langle \phi_-|(\hat{\mathbf{e}} \cdot \boldsymbol{\sigma})(\hat{\mathbf{n}} \cdot \boldsymbol{\sigma})|\phi_-\rangle = -\hat{\mathbf{e}} \cdot \hat{\mathbf{n}}$.

2. Alice and Bob are using the B92 protocol with states $|\psi_0\rangle$ and $|\psi_1\rangle$. Eve captures the qubit going from Alice to Bob and entangles it with an ancilla qubit and then sends Alice's original qubit on to Bob. Eve wants to measure the ancilla qubit to gain information about the qubit Alice sent to Bob. In particular, Eve uses the unitary entangling operation U to perform

$$U|\psi_0\rangle_A|0\rangle_E = |\psi_0\rangle_A|v_{00}\rangle_E + |\psi_0^\perp\rangle_A|v_{01}\rangle_E$$
$$U|\psi_1\rangle_A|0\rangle_E = |\psi_1\rangle_A|v_{11}\rangle_E + |\psi_1^\perp\rangle_A|v_{10}\rangle_E,$$

where $\langle \psi_j|\psi_j^\perp\rangle = 0$, for $j = 0,1$, and the vectors $|v_{jk}\rangle$ for $j,k = 0,1$ are not necessarily normalized. Show that if Eve is to create no errors she will gain no information about Alice's qubit.

3. Another way to do quantum secret sharing is to use the GHZ state $|\Psi\rangle_{abc} = (1/\sqrt{2})(|000\rangle_{abc} + |111\rangle_{abc})$. Define the x and y bases as $|\pm x\rangle = (1/\sqrt{2})(|0\rangle \pm |1\rangle)$ and $|\pm y\rangle = (1/\sqrt{2})(|0\rangle \pm i|1\rangle)$. Alice, Bob, and Charlie each have one of the qubits in the GHZ state. Show that if Alice and Bob both measure in the same basis, and Charlie measures in the x basis, then if Bob and Charlie communicate their measurement results to each other, they can determine the result of Alice's measurement. In addition, show that if Alice and Bob measure in different bases, and Charlie measures in the y basis, then, again, if Bob and Charlie communicate their measurement results to each other, they can determine the result of Alice's measurement. Therefore, Alice can establish a joint key with Bob and Charlie, but Bob and Charlie have to cooperate to obtain it.

References

1. C.H. Bennett, Quantum cryptography using any two nonorthogonal states. Phys. Rev. Lett. **68**, 3121 (1992)
2. C.H. Bennett, G. Brassard, *Quantum Cryptography: Public Key Distribution and Coin Tossing*. In Proceedings of IEEE International Conference on Computers, Systems, and Signal Processing, Bangalore, India, December 1984 (IEEE, New York, 1984), p. 175
3. A. Ekert, Quantum cryptography based on Bell's theorem. Phys. Rev. Lett. **67**, 661 (1991)
4. M. Hillery, V. Bužek, A. Berthiaume, Quantum secret sharing. Phys. Rev. A **59**, 1829 (1999)
5. R. Cleve, D. Gottesman, Hoi-Kwong Lo, How to share a quantum secret. Phys. Rev. Lett. **83**, 648 (1999)
6. J. Mimih, M. Hillery, Unambiguous discrimination of special sets of multipartite states using local measurements and classical communication. Phys. Rev. A **71**, 012329 (2005)
7. N. Gisin, G. Ribordy, W. Tittel, H. Zbinden, Quantum cryptography. Rev. Mod. Phys. **74**, 145 (2002)

Chapter 7
Quantum Algorithms

In this chapter we shall look at a number of quantum algorithms. We are going to compare their performance, in terms of number of steps, to classical algorithms that accomplish the same task.

7.1 The Deutsch–Jozsa Algorithm

We shall start with a generalization of the Deutsch algorithm, known as the Deutsch–Jozsa algorithm. The problem can be stated as follows: Given a Boolean function on n-digit binary numbers, $f : \{0,1\}^n \to \{0,1\}$, which is promised to be either constant or balanced, determine which.

Classically, $2^{(n-1)} + 1$ function evaluations are necessary in the worst case scenario. There is a quantum algorithm that requires only one function evaluation. The corresponding quantum circuit is shown in Fig. 7.1.

In order to understand how this circuit works we shall analyze the state of the $n + 1$ qubit system at each step, i.e., at the input, after the first set of Hadamard gates, after the f-Controlled-NOT gate, and after the second set of Hadamard gates which constitutes the output state generated by the circuit.

Since Hadamard gates are placed before and after the f-CNOT gate, we begin by analyzing the action of a set of Hadamard gates on a general binary number state which, in turn, is defined as follows. Let $x = x_{n-1}x_{n-2}\ldots x_0$, where $x_j \in \{0,1\}$, stand for an n-digit binary number. Then the binary number state $|x\rangle$ of an n qubit system in the computational basis is given by $|x\rangle = |x_{n-1}\rangle \otimes |x_{n-2}\rangle \ldots \otimes |x_0\rangle$. The action of the Hadamard gate on a single qubit in the computational basis state can be summarized as $H|x_j\rangle \to \frac{1}{\sqrt{2}}(|0\rangle + (-1)^{x_j}|1\rangle)$, so

J.A. Bergou and M. Hillery, *Introduction to the Theory of Quantum Information Processing*, Graduate Texts in Physics, DOI 10.1007/978-1-4614-7092-2_7, © Springer Science+Business Media New York 2013

Fig. 7.1 Quantum circuit for
the Deutsch–Jozsa problem

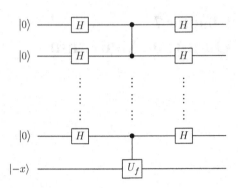

$$\Pi_{j=0}^{n-1}|x_j\rangle \rightarrow \left(\frac{1}{\sqrt{2}}\right)^n \prod_{j=0}^{n-1}(|0\rangle + (-1)^{x_j}|1\rangle)$$

$$= \left(\frac{1}{\sqrt{2}}\right)^n \sum_{z=0}^{2^n-1}\left(\prod_{j=\text{ such that } z_j=1}(-1)^{x_j}\right)|z\rangle$$

$$= \left(\frac{1}{\sqrt{2}}\right)^n \sum_{z=0}^{2^n-1}\left(\prod_{j=0}^{n-1}(-1)^{x_j z_j}\right)|z\rangle. \tag{7.1}$$

Notice that $\prod_{j=0}^{n-1}(-1)^{x_j z_j} = (-1)^{\sum_{j=0}^{n-1}x_j z_j} = (-1)^{\left[\sum_{j=0}^{n-1}x_j z_j \bmod 2\right]}$. Let us define the
dot product as $x \cdot z \equiv \sum_{j=0}^{n-1}x_j z_j \bmod 2$, then

$$|x\rangle \rightarrow \left(\frac{1}{2}\right)^{n/2}\sum_{z=0}^{2^n-1}(-1)^{x \cdot z}|z\rangle. \tag{7.2}$$

In other words, a given n-digit binary number state is turned into an equally
weighted superposition of all 2^n binary number states of the n qubits and the sign of
each term is determined by the parity of the dot product between the given state and
binary state in the term.

If now we apply this to the input state of the n control qubits, $|\psi_{in}\rangle = |0\rangle$, we
obtain the state of the n-qubit system after the first set of Hadamards as

$$|\psi_1\rangle = \left(\frac{1}{2}\right)^{n/2}\sum_{z=0}^{2^n-1}|z\rangle. \tag{7.3}$$

Next we analyze the action of the f-CNOT gate on this state. To this end we note
that the f-Controlled-NOT gate acts as $|x\rangle|y\rangle \rightarrow |x\rangle|y+f(x) \bmod 2\rangle$ where $|x\rangle$ is the
state of the n control qubits and $|y\rangle$ is the state of the single target qubit. Therefore,

$$|x\rangle \otimes \frac{1}{\sqrt{2}}(|0\rangle - |1\rangle) \rightarrow |x\rangle \otimes (|f(x)\rangle - |1 + f(x)\rangle)$$

$$= (-1)^{f(x)}|x\rangle \otimes \frac{1}{\sqrt{2}}(|0\rangle - |1\rangle). \tag{7.4}$$

Combining this with $|\psi_1\rangle$ in Eq. (7.3), we obtain the state after the f-CNOT gate as

$$|\psi_2\rangle \otimes \frac{1}{\sqrt{2}}(|0\rangle - |1\rangle) = \left(\frac{1}{2}\right)^{(n+1)/2} \sum_{x=0}^{2^n-1} (-1)^{f(x)}|x\rangle \otimes (|0\rangle - |1\rangle). \tag{7.5}$$

Finally, applying Eq. (7.2) to this state yields the output state after the final set of Hadamard gates as

$$|\psi_{in}\rangle \otimes \frac{1}{\sqrt{2}}(|0\rangle - |1\rangle) \rightarrow \left(\frac{1}{2}\right)^{n+1/2} \sum_{x,z=0}^{2^n-1} (-1)^{f(x)+x\cdot z}|z\rangle \otimes (|0\rangle - |1\rangle)$$

$$= |\psi_{out}\rangle \otimes \frac{1}{\sqrt{2}}(|0\rangle - |1\rangle). \tag{7.6}$$

The amplitude of the initial state, $|\psi_{in}\rangle = |0\rangle$, in the output state is easily obtained as $\langle 0|\psi_{out}\rangle = \left(\frac{1}{2}\right)^n \sum_{x=0}^{2^n-1}(-1)^{f(x)}$, and

$$\langle 0|\psi_{out}\rangle = \begin{cases} 0 & \text{if f(x) balanced,} \\ (-1)^{f(0)} & \text{if f(x) constant} \rightarrow |\psi_{out}\rangle = (-1)^{f(0)}|0\rangle. \end{cases} \tag{7.7}$$

Therefore measuring each of the n output qubits we have with certainty that:

1. f(x) = constant if we find all qubits in their 0 state.
2. f(x) = balanced if not all of them are found in their 0 state.

Note that this is accomplished with only one function evaluation.

7.2 The Bernstein–Vazirani Algorithm

We can use the Deutsch-Jozsa circuit to solve another problem, which is due to Bernstein and Vazirani. Suppose

$$f(x) = a \cdot x + b \,(\text{mod } 2), \tag{7.8}$$

where $a \in \{0,1\}^n$ and $b \in \{0,1\}$. Our goal is to determine a (we do not know a or b). Classically, because a contains n bits of information, we are going to have to evaluate $f(x)$ n times at least. One method is to evaluate it for $x = 0$, giving b, and then for $x_j = 0\ldots010\ldots0$, where the 1 is in the jth place, for $j = 1,\ldots,n$.

With this $f(x)$ our state at the output of the quantum circuit is

$$|\Psi_{out}\rangle = \left(\frac{1}{2}\right)^n \sum_{x,y=0}^{2^n-1} (-1)^b (-1)^{x\cdot(a+y)} |y\rangle. \tag{7.9}$$

where $(a+y)$ in the exponent stands for bitwise addition.

We show that $\sum_{x=0}^{2^n-1} (-1)^{x\cdot z} = 0$ unless $z \in \{0,1\}^n = 0$. This can be seen as follows. First, we rewrite the sum as

$$\sum_{x=0}^{2^n-1} (-1)^{x\cdot z} = \sum_{x=0}^{2^n-1} \prod_{j=0}^{n-1} (-1)^{x_j z_j} = \sum_{x_{n-1}=0}^{1} \cdots \sum_{x_0=0}^{1} \prod_{j=0}^{n-1} (-1)^{x_j z_j}. \tag{7.10}$$

Suppose now $z_k = 1$, then

$$\sum_{x} (-1)^{x\cdot z} = \sum_{x_{n-1}=0}^{1} \cdots \sum_{x_{k+1}=0}^{1} \sum_{x_{k-1}=0}^{1} \cdots \sum_{x_0=0}^{1} \prod_{j=0, j\neq k}^{n-1} (-1)^{x_j z_j} (1+(-1)) = 0 , \tag{7.11}$$

where the last two terms in the bracket arise from $x_k = 0$ yielding the $+1$ and $x_k = 1$ yielding the (-1). Therefore,

$$\sum_{x} (-1)^{x\cdot z} = 2^n \delta_{z,0},$$

and

$$|\Psi_{out}\rangle = (-1)^b |a\rangle, \tag{7.12}$$

so that measuring the n output qubits in $|\Psi_{out}\rangle$ gives us a with only one function evaluation.

7.3 Quantum Search: The Grover Algorithm

Typically, Grover's problem can be stated as the search for one marked entry in an unsorted database. Mathematically, it can be formulated as the following problem. Let $f(x) = 0$ or 1 where x is an n bit binary number. In particular

$$f(x) = \begin{cases} 1 & \text{if } x = x_0, \\ 0 & \text{if } x \neq x_0. \end{cases} \tag{7.13}$$

x_0 is unknown and we would like to find it. The search is schematically depicted in Fig. 7.2.

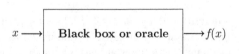

Fig. 7.2 Scheme of the search problem

The central question is: How many function evaluations are necessary? Classically, if $N = 2^n$, then $\mathcal{O}(N)$ evaluations are necessary. On a quantum computer it can be done with $\mathcal{O}(\sqrt{N})$ evaluations. (Our treatment is taken from R. Jozsa, quant-ph/990121.)

To this end, define the following operators:

$$U_f|x\rangle = (-1)^{f(x)}|x\rangle = (-1)^{\delta_{x,x_0}}|x\rangle,$$

$$U_0|x\rangle = (-1)^{\delta_{x,0}}|x\rangle = (I - 2|0\rangle\langle 0|)|x\rangle,$$

$$U_H = (H)^{\otimes n}. \tag{7.14}$$

An alternative form of U_f is given by $U_f = I - 2|x_0\rangle\langle x_0|$ and we already know the circuit for this operator.

Grover's algorithm consists in applying the operator $Q = -U_H U_0 U_H U_f$ to the initial state, $|w_0\rangle = U_H|0\rangle = \frac{1}{\sqrt{N}}\sum_{x=0}^{N-1}|x\rangle$, $\mathcal{O}(\sqrt{N})$ times and then measuring the state in the computational basis. The answer will be, with probability greater than $\frac{1}{2}$, x_0 (actually, with a probability close to 1).

How does this work? First define $U_{w_0} = U_H U_0 U_H = I - 2|w_0\rangle\langle w_0|$ and $S = \text{span}\{|w_0\rangle, |x_0\rangle\}$ which is a two-dimensional subspace. For any $|\psi\rangle = c_1|w_0\rangle + c_2|x_0\rangle \in S$, we have

$$Q|\psi\rangle = -U_{w_0}U_f(c_1|w_0\rangle + c_2|x_0\rangle) = -U_{w_0}[c_1(|w_0\rangle - \frac{2}{\sqrt{N}}|x_0\rangle) - c_2|x_0\rangle]$$

$$= c_1|w_0\rangle + \left(\frac{2}{\sqrt{N}} + c_2\right)(|x_0\rangle - \frac{2}{\sqrt{N}}|x_0\rangle) \in S, \tag{7.15}$$

so that Q maps S into itself. Therefore, all of the action in Grover's algorithm takes place in a 2D subspace. Note also that if c_1 and c_2 are real so are the coefficients of $|w_0\rangle$ and $|x_0\rangle$. As we start by applying Q to $|w_0\rangle$, we actually need only to consider $S' = \{c_1|w_0\rangle + c_2|x_0\rangle | c_1, c_2 \text{ real}\}$, i.e. S' is a real 2D subspace.

Now look at Q more closely. The operator U_f in S' is just a reflection about the line parallel to $|x_0^{\perp}\rangle$. Note:

$$|x_0^{\perp}\rangle = (|w_0\rangle + |x_0\rangle\langle x_0|w_0\rangle)/(1 - |\langle x_0|w_0\rangle|^2)^{1/2},$$

and

$$|w_0^{\perp}\rangle = (|x_0\rangle + |w_0\rangle\langle w_0|x_0\rangle)/(1 - |\langle x_0|w_0\rangle|^2)^{1/2}.$$

We also have that, in the subspace S', $|w_0\rangle\langle w_0| + |w_0^{\perp}\rangle\langle w_0^{\perp}| = I$. From here, it follows that $-U_f = -(I - 2|w_0\rangle\langle w_0|) = I - 2|w_0^{\perp}\rangle\langle w_0^{\perp}| = U_{w_0^{\perp}}$ and this is a reflection about the line through $|w_0\rangle$. Therefore, Q corresponds to two consecutive reflections,

$$Q = U_{w_0^{\perp}}U_f = (\text{reflection about } w_0)(\text{reflection about } x_0^{\perp}). \tag{7.16}$$

The geometry of the two reflections is shown in Fig. 7.3.

Fig. 7.3 Geometry
associated with the Grover
search algorithm

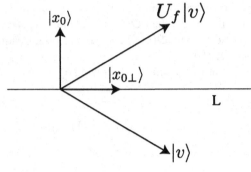

Fig. 7.4 Two subsequent
reflections of $\mathbf{v}_1$, R_2R_1, first
through M_1 followed by
another through M_2,
correspond to an effective
rotation by the angle 2α

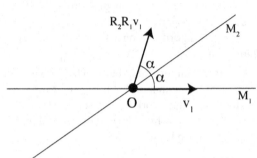

Theorem 1 *Let M_1 and M_2 be two mirror lines in Euclidean plane $\mathbb{R}^2$ intersecting at point O and α be the angle from M_1 to M_2. The operation of reflection through M_1 followed by reflection through M_2 is a rotation by 2α about O.*

Proof. The proof uses pictorial but nevertheless rigorous arguments. Let M_1 be parallel to $\mathbf{v}_1$ and M_2 parallel to $\mathbf{v}_2$. If the theorem holds for $\mathbf{v}_1$ and $\mathbf{v}_2$ it holds for any superposition of them, hence for any vector. Let R_1 be the reflection through M_1 and R_2 be the reflection through M_2. We will now separately study what happens to $\mathbf{v}_1$ and $\mathbf{v}_2$, as a result of these two reflections. First, look at $\mathbf{v}_1$. A reflection of $\mathbf{v}_1$ through $M_1 = \mathbf{v}_1$ maps $\mathbf{v}_1$ onto itself. A subsequent reflection of $\mathbf{v}_1$ through $M_2 = \mathbf{v}_2$ corresponds to an effective rotation of $\mathbf{v}_1$ by the angle 2α in the counterclockwise direction. The situation is shown in Fig. 7.4.

Next, look at $\mathbf{v}_2$. The situation is shown in Fig. 7.5. A reflection of $\mathbf{v}_2$ through $M_1 = \mathbf{v}_1$ rotates $\mathbf{v}_2$ by the angle 2α in the clockwise direction. A subsequent reflection of $R_1\mathbf{v}_2$ around $M_2 = \mathbf{v}_2$ corresponds to an effective rotation of $\mathbf{v}_2$ by the angle 2α relative to its original orientation.

Therefore, Q is a rotation in S' by angle 2α, where α is the angle between $|w_0\rangle$ and $|x_0^\perp\rangle$. Furthermore,

$$\cos\alpha = \langle w_0 | x_0^\perp \rangle = \left(1 - \frac{1}{\sqrt{N}}\right)^{\frac{1}{2}},$$

Fig. 7.5 Two subsequent reflections of v_2, $R_2 R_1$, first through M_1 followed by another through M_2, correspond to an effective rotation by the angle 2α

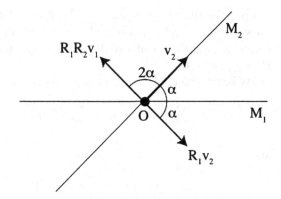

and

$$\sin \alpha = \langle w_0 | x_0 \rangle = \left(1 - \cos^2 \alpha\right)^{\frac{1}{2}} = \frac{1}{\sqrt{N}}.$$

$\square$

Starting with the state (in the $|x_0\rangle$, $|x_{0\perp}\rangle$ basis)

$$|w_0\rangle = |x_0\rangle |\langle x_0 | w_0\rangle + |x_{0\perp}\rangle \langle x_{0\perp} | w_0\rangle$$
$$= \sin \alpha |x_0\rangle + \cos \alpha |x_{0\perp}\rangle, \tag{7.17}$$

we have

$$Q^n = |w_0\rangle = \sin \alpha_n |x_0\rangle + \cos \alpha_n |x_{0\perp}\rangle, \tag{7.18}$$

where $\alpha_n = (2n+1)\alpha$. We want to choose n so that α_n is close to $\pi/2$. For large N, $\alpha \cong \frac{1}{\sqrt{N}}$ so we want $(2n+1)\frac{1}{\sqrt{N}} \cong \frac{\pi}{2}$. Therefore, $n = $ closest integer to $\frac{\pi}{4}\sqrt{N} - \frac{1}{2}$. Let us call this value $\bar{n}$. Then the probability of measuring $x_0 = |\langle x_0 | Q^{\bar{n}} | w_0 \rangle|^2 = \sin^2 \alpha_{\bar{n}} \simeq 1$ and the probability of measuring $x \neq x_0 = |\langle x | Q^{\bar{n}} | w_0 \rangle|^2 = \cos^2 \alpha_{\bar{n}} = \mathcal{O}(\frac{1}{N^2})$.

Is the speed up from N calls of the black box that evaluates the function to $\sqrt{N}$ calls the best we can do? The answer is yes. To show this assume the algorithm works by alternating function calls with unitary evolution. The function call is described by $U_x = I - 2|x\rangle\langle x|$; hence

$$U_x |y\rangle = \begin{cases} |y\rangle & \text{if } y \neq x, \\ -|x\rangle & \text{if } y = x. \end{cases} \tag{7.19}$$

After k function calls the state of the system is

$$|\psi_k^x\rangle = U_k U_x U_{k-1} U_x \ldots U_1 U_x |\psi_{in}\rangle. \tag{7.20}$$

Then the strategy is as follows. Compare $|\psi_k^x\rangle to |\psi_k\rangle = U_k U_{k-1} \dots U_1 |\psi_{in}\rangle$ to show that if the probability of finding x is great, in particular, if $|\langle x|\psi_k^x\rangle > \frac{1}{2}$, then k must be of the order of $\sqrt{N}$. In particular, we find upper and lower bounds for $D_k = \sum_x \|\psi_k^x - \psi_k\|^2$.

We begin with establishing the upper bound. First, note that

$$D_{k+1} = \sum_x \|U_x \psi_k^x - \psi_k\|^2 = \sum_x \|U_x(\psi_k^x - \psi_k) + (U_x - I)\psi_k\|^2, \qquad (7.21)$$

and

$$D_{k+1} \leq \sum_x (\|\psi_k^x - \psi_k\| + \|(U_x - I)\psi_k\|)^2$$

$$= \sum_x (\|\psi_k^x - \psi_k\|^2 + 4\|\psi_k^x - \psi_k\| |\langle x|\psi_k\rangle| + |\langle x|\psi_k\rangle|^2)$$

$$\leq D_k + 4 \left(\sum_x \|\psi_k^x - \psi_k\|^2 \right)^{1/2} \left(\sum_x |\langle x|\psi_k\rangle|^2 \right)^{1/2} + 4$$

$$\leq D_k + 4\sqrt{D_k} + 4. \qquad (7.22)$$

Now, we will use this result and induction to show $D_k \leq 4k^2$. First, we have $D_0 = 0$. Now

$$D_1 = \sum_x \|U_1 U_x \psi_{in} - U_1 \psi_{in}\|^2 = \sum_x \|U_x \psi_{in} - \psi_{in}\|^2, \qquad (7.23)$$

but $(U_x - I)|\psi_{in}\rangle = -2|x\rangle\langle x|\psi_{in}\rangle$, so we have that

$$D_1 = 4 \sum_x |\langle x|\psi_{in}\rangle|^2 = 4. \qquad (7.24)$$

So, it is clearly true that $D_k \leq 4k^2$ for $k = 0, 1$. Now, assuming it is true for k

$$D_{k+1} \leq D_k + 4\sqrt{D_k} + 4 \leq 4k^2 + 8k + 4 = 4(k+1)^2. \qquad (7.25)$$

Therefore, $D_k \leq 4k^2$.

Next we establish the lower bound. To this end, let us define $Q_x = I - |x\rangle\langle x|$, then we have that

$$\|\psi_k^x - \psi_k\|^2 = \||x\rangle(\langle x|\psi_k^x\rangle - \langle x|\psi_k\rangle) + Q_x(\psi_k^x - \psi_k)\|^2$$

$$= |\langle x|\psi_k^x\rangle - \langle x|\psi_k\rangle|^2 + \|Q_x(\psi_k^x - \psi_k)\|^2$$

$$\geq |\langle x|\psi_k^x\rangle|^2 + |\langle x|\psi_k\rangle|^2 - 2|\langle x|\psi_k^x\rangle| \cdot |\langle x|\psi_k\rangle|$$

$$+ \|Q_x \psi_k^x\|^2 + \|Q_x \psi_k\|^2 - 2|\langle Q_x \psi_k^x|\psi_k\rangle|$$

$$\geq 2 - 2|\langle x|\psi_k\rangle| - 2\|Q_x \psi_k^x\|. \qquad (7.26)$$

Now suppose we assume that after k steps, when we measure the state $|\psi_k^x\rangle$, our probability of finding x is greater than $1/2$, i.e., $|\langle x|\psi_k^x\rangle|^2 > 1/2$. We also have that $|\langle x|\psi_k^x\rangle|^2 + \|Q_x\psi_k^x\|^2 = 1$ from which $\|Q_x\psi_k^x\|^2 \leq 1/2$ follows, so $\|\psi_k^x - \psi_k\|^2 \geq 2 - 2|\langle x|\psi_k\rangle| - \sqrt{2}$ and

$$D_k \geq \sum_x (2 - 2|\langle x|\psi_k\rangle| - \sqrt{2})$$

$$\geq N(2 - \sqrt{2}) - 2\left(\sum_x 1^2\right)^{1/2}\left(\sum_x |\langle x|\psi_k\rangle|^2\right)^{1/2}$$

$$\geq N(2 - \sqrt{2}) - 2\sqrt{N}. \tag{7.27}$$

Putting the bounds together gives

$$4k^2 \geq N(2 - \sqrt{2}) - 2\sqrt{N} \tag{7.28}$$

and from here

$$k \geq \frac{(2 - \sqrt{2})^{1/2}}{2}\sqrt{N}\left(1 - \frac{2}{\sqrt{N}}\frac{1}{2 - \sqrt{2}}\right)^{1/2} \tag{7.29}$$

follows. So, reducing the number of function calls from N to $\sqrt{N}$ is indeed the best we can do and the Grover search algorithm is optimal.

7.4 Period Finding: Simon's Algorithm

Now we take a look at Simon's algorithm which is a simple period-finding algorithm. A more sophisticated version is one of the major components of Shor's factoring algorithm.

Consider a function $F : \mathbb{Z}_2^{\otimes n} \to \mathbb{Z}_2^{\otimes n}$ which is $2 \to 1$. In particular

$$f(x) = f(y) \quad \text{iff } y = x \oplus \xi \text{ where } x, y, \xi \in \mathbb{Z}_2^{\otimes n}. \tag{7.30}$$

Here $\oplus$ stands for component-wise mod 2 addition, i.e., for $w, z \in \mathbb{Z}_2^{\otimes n}$ we have that $w \oplus z = (w_1 + z_1 \ (mod\ 2), \ldots, w_n + z_n \ (mod\ 2))$ and ξ is fixed. The object is to find ξ with only $poly(n)$ function evaluations.

Start with $|0\ldots0\rangle$ and apply a Hadamard gate to each qubit to get $2^{-n/2}\sum_x |x\rangle$. Now apply U_f, which has the following action

$$U_f|x\rangle|y\rangle = |x\rangle|y \oplus f(x)\rangle. \tag{7.31}$$

So

$$U_f\left(\frac{1}{2^{n/2}}\sum_x |x\rangle|0\rangle\right) = \frac{1}{2^{n/2}}\sum_x |x\rangle|f(x)\rangle. \tag{7.32}$$

Now measure the second register. This gives some result, x_0, and leaves the first in the state $\frac{1}{\sqrt{2}}(|x_0\rangle + |x_0 \oplus \xi\rangle)$, where x_0 is random. This randomness makes measuring the above state useless, if we want to determine ξ. Instead, apply $H^{\otimes n}$ to the state. This gives us

$$\frac{1}{2^{(n+1)/2}} \sum_y \left[(-1)^{x_0 \cdot y} + (-1)^{(x_0 \oplus \xi) \cdot y} \right] |y\rangle$$

$$= \frac{1}{2^{(n+1)/2}} \sum_y (-1)^{x_0 \cdot y} \left[1 + (-1)^{\xi \cdot y} \right] |y\rangle$$

$$= \frac{1}{2^{(n-1)/2}} \sum_{\{y | y \cdot \xi = 0\}} (-1)^{x_0 \cdot y} |y\rangle. \tag{7.33}$$

Now measure this state. We get some value of y, call it y_1, such that $y_1 \cdot \xi = 0$. With $\mathcal{O}(n)$ iterations of this procedure, we obtain n independent equations of the form $y_j \cdot \xi = 0$, $j = 1, \ldots, n$, and we can solve this linear system to determine ξ.

7.5   Quantum Fourier Transform and Phase Estimation

The quantum Fourier transform is a component of a number of quantum algorithms, the Shor factoring algorithm in particular. We will not treat the Shor algorithm in this book, as it has been covered extensively in many other places. We will show instead how the quantum Fourier transform can be used to find an unknown eigenvalue of a unitary transformation.

Let $|a\rangle$ be a member of the computational basis in the m-qubit Hilbert space. The m-bit binary number a can be expressed as $a = 2^{m-1}a_1 + 2^{m-1}a_2 + \cdots + 2^0 a_m$, where each of the a_j is either 0 or 1. The quantum Fourier transform, U_F, takes $|a\rangle$ into the state

$$U_F |a\rangle = \frac{1}{2^m} \sum_{y=0}^{2^m - 1} e^{2\pi i a \cdot y / 2^m} |y\rangle. \tag{7.34}$$

The inverse transformation is given by

$$U_F^{-1} |a\rangle = \frac{1}{2^m} \sum_{y=0}^{2^m - 1} e^{-2\pi i a \cdot y / 2^m} |y\rangle. \tag{7.35}$$

This transformation can be implemented efficiently using only one- and two-qubit gates.

Now let us see how we can use the quantum Fourier transform to estimate an unknown eigenvalue. The circuit is shown in Fig. 7.6.

Suppose that we have the unitary operator U, where $U|\psi\rangle = \exp(2\pi i \phi)|\psi\rangle$ and $0 \leq \phi < 1$. We are given one copy of $|\psi\rangle$ and gates that perform Controlled-U^k

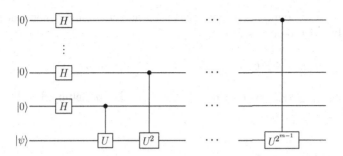

Fig. 7.6 Quantum circuit for phase estimation

operations for $k = 1, 2, 2^2, \ldots 2^{m-1}$. We want to find ϕ, which we do not know, to m-bit accuracy. We start with each of the qubits in the m control lines in the state $(|0\rangle + |1\rangle)/\sqrt{2}$, so that the initial state of our computation is

$$2^{-m/2} \left[\prod_{j=0}^{m-1} (|0\rangle_j + |1\rangle_j) \right] \otimes |\psi\rangle. \tag{7.36}$$

We now apply the Controlled-U^{2^j} gates, the control being the jth qubit and the target being the system in the state $|\psi\rangle$. This results in the state

$$2^{-m/2} \left[\prod_{j=0}^{m-1} (|0\rangle_j + e^{2\pi i 2^j \phi} |1\rangle_j) \right] \otimes |\psi\rangle = 2^{-m/2} \sum_{y=0}^{2^m-1} e^{2\pi i \phi y} |y\rangle \otimes |\psi\rangle, \tag{7.37}$$

where $|y\rangle$ is an m-qubit computational basis state. Now if ϕ is of the form $a/2^m$, where a is an m-digit binary number, we can simply apply the inverse quantum Fourier transform to the above state, and the result will be $|a\rangle$. We will then have learned ϕ.

Now let us see what happens if ϕ is not of the form $a/2^m$. Let $\phi = (a/2^m) + \delta$, where a is the closest m-bit binary number to $2^m \phi$. This implies that $0 < |\delta| \leq 2^{-(m+1)}$. We now apply the inverse Fourier transform to the state in Eq. (7.37) yielding

$$2^{-m} \sum_{y=0}^{2^m-1} \sum_{x=0}^{2^m-1} e^{-2\pi i x \cdot y/2^m} e^{2\pi i \phi y} |x\rangle = 2^{-m} \sum_{y=0}^{2^m-1} \sum_{x=0}^{2^m-1} e^{2\pi i (a-x) \cdot y/2^m} e^{2\pi i \delta y} |x\rangle, \tag{7.38}$$

where we have dropped $|\psi\rangle$ since it is not entangled with the rest of the state and plays no further role. Now let us look at the coefficient of the state $|a\rangle$ in the above equation. It is given by

$$2^{-m} \sum_{y=0}^{2^m-1} e^{2\pi i \delta y} = 2^{-m} \left(\frac{1 - e^{2\pi i \delta 2^m}}{1 - e^{2\pi i \delta}} \right). \tag{7.39}$$

We now want to bound the magnitudes of the numerator and denominator of this fraction. In order to do so we note that

$$|1 - e^{i\theta}| = \sqrt{2}(1 - \cos\theta)^{1/2} = 2\sin(\theta/2). \tag{7.40}$$

Now for $0 \leq \beta \leq \pi/2$, we have that $(2/\pi)\beta \leq \sin\beta \leq \beta$. Setting $\beta = \theta/2$ we have that for $0 \leq \theta \leq \pi$,

$$\frac{2\theta}{\pi} \leq |1 - e^{i\theta}| \leq \theta. \tag{7.41}$$

Note that because $|1 - e^{i\theta}| = |1 - e^{-i\theta}|$, the above inequalities can be modified to hold in the range $-\pi \leq \theta \leq \pi$ by inserting absolute value signs appropriately

$$\frac{2|\theta|}{\pi} \leq |1 - e^{i\theta}| \leq |\theta|. \tag{7.42}$$

Now because $|\delta| \leq 1/2^{m+1}$, we have that $2\pi\delta 2^m \leq \pi$ and, therefore, $|1 - e^{2\pi i\delta 2^m}| \geq 4\delta 2^m$, and we also have that $|1 - e^{2\pi i\delta}| \leq 2\pi\delta$. This implies that the probability of obtaining the state $|a\rangle$ when measuring the output state of the circuit is

$$2^{-2m} \left| \frac{1 - e^{2\pi i\delta 2^m}}{1 - e^{2\pi i\delta}} \right|^2 \geq 2^{-2m} \left(\frac{4\delta 2^m}{2\pi\delta} \right)^2 = \frac{4}{\pi^2}. \tag{7.43}$$

Therefore, the probability of obtaining the best m-bit approximation to ϕ is $(4/\pi^2) = 0.4$. A more detailed analysis shows that the probability of getting an error greater than $k/2^m$ is less than $1/(2k-1)$.

One possible use for this algorithm is related to the Grover search. Suppose we are given a black box Boolean function that is of one of two types. There is either one input, x_0, which we do not know, for which $f(x_0) = 1$, with all other inputs, $x \neq x_0$ yielding $f(x) = 0$, or all inputs yield $f(x) = 0$. We would like to find which type of black box function we have. One approach is to run the Grover algorithm and see if we get the same answer almost all of the time. If so, we have the first kind of black box. If we get different answers each time, then we have the second type. A second approach is to use the phase estimation algorithm. The operator $Q = U_{w_0^\perp} U_f$ has different eigenvalues for the two different types of oracles. In the case that all inputs yield $f(x) = 0$, we have that $U_f = I$, which implies that $Q = U_{w_0^\perp}$. In that case, Q is just a reflection, so that its eigenvalues are just ± 1. In particular, the state $|w_0\rangle$ is an eigenstate with eigenvalue 1. If one of the inputs yields $f(x_0) = 1$, then, in the subspace S', Q can be expressed as a 2×2 matrix in the $\{|w_0\rangle, |w_0^\perp\rangle\}$ basis

$$Q = \begin{pmatrix} \cos 2\alpha & -\sin 2\alpha \\ \sin 2\alpha & \cos 2\alpha \end{pmatrix}, \tag{7.44}$$

where α is the angle between $|w_0\rangle$ and $|x_0^\perp\rangle$ and is $O(N^{-1/2})$. This matrix has eigenvalues $e^{\pm 2i\alpha}$, and the eigenstates are $|\alpha_\pm\rangle = (|w_0\rangle \mp i|w_0^\perp\rangle)/\sqrt{2}$. Now suppose that N, the number of possible inputs to our Boolean function, is $N = 2^n$. In order to discriminate between the two types of oracles, we need to determine the eigenvalues of Q to $O(2^{-n/2})$, because $1 - e^{2i\alpha}$ is of this order. We then make use of the phase estimation algorithm with $m > n/2$ and an input state into the target qubits of the Controlled-Q^{2^j} gates of $|w_0\rangle$. Now $|w_0\rangle$ is not an eigenstate of Q, but it is the sum of two eigenstates $|w_0\rangle = (|\alpha_+\rangle + |\alpha_-\rangle)/\sqrt{2}$. The output of the phase estimation circuit will be approximately of the form $(|a_+\rangle|\alpha_+\rangle + |a_-\rangle|\alpha_-\rangle)/\sqrt{2}$ where $a_+/2^m$ is a good estimate of $\alpha/2\pi$ and $a_-/2^m$ is a good estimate of $(2\pi - \alpha)/2\pi$. If we simply measure the first m qubits of the output state in the computational basis, we will obtain, with equal probability, an estimate of either a_+ or a_-. If either one of these is different from zero, then we know that there is an x_0 such that $f(x_0) = 1$.

The procedure we have just outlined is most useful when there is more than one value of x such that $f(x) = 1$, and we want to find out how many values of x satisfying this condition there are. This is a procedure known as quantum counting. In that case the eigenvalues of Q depend on the number of solutions, and by estimating the eigenvalues we can determine that number.

7.6 Quantum Walks

Finding new quantum algorithms has not been easy, and one approach one might try to find new ones is to see if there are particular mathematical structures that have proved useful in classical algorithms and then try to generalize them to the quantum realm. One area in which this approach has been fruitful is in algorithms based on random walks. There are a number of classical algorithms based on random walks, and we shall present an example of one shortly. It has been possible to define a quantum version of a random walk, known as a quantum walk, and there are now new quantum algorithms that are based on quantum walks. In this section we will describe what a quantum walk is and some of the things they can do.

The simplest example of a classical random walk is one on a line. The walk starts at a point, which we shall call the origin. The walker then flips an unbiased coin. If it comes up heads, he takes one step to the right, if tails, one step to the left (all steps are the same length). This process is repeated for the desired number of steps, n. The result can be described by a probability distribution, $p(x;n)$, which is the probability of being at position x after n steps. The position is measured in units of step length and is positive to the right of the origin (which is $x = 0$) and negative to the left. For example, for a walk of two steps, the only possible final positions are $x = -2, 0, 2$ and we find that $p(-2;2) = p(2;2) = 1/4$ and $p(0;2) = 1/2$.

It is also possible to perform random walks on more general structures known as graphs. A graph consists of a set of vertices, V, and a set of edges, E. Each edge connects two of the vertices, and an edge is labeled by an unordered pair of vertices,

which are just the vertices connected by that edge. In general, not all of the vertices
will be connected by an edge. A graph in which each pair of vertices is connected
by an edge is known as a complete graph, and if there are N vertices, there will
be $N(N-1)/2$ edges in a complete graph. In order to perform a random walk on
a graph, we choose one vertex on which to start. For the first step, we see which
vertices are connected to the vertex we are on by an edge, and then we randomly
choose one of them, each having the same probability, and then move to that vertex.
So, for example, if our starting vertex is connected to three other vertices, then we
would end up on each of those vertices with a probability of $1/3$. We then repeat
this process for the new vertex in order to make the second step and keep repeating
it for as many steps as we wish.

A simple example of an algorithm based on a random walk is one that determines
whether two vertices in a graph are connected or not. In order to determine whether
there is a path connecting a specified vertex u to another specified vertex v, we can
start a walker at u, execute a random walk for a certain number of steps, and see
after each step whether we have reached v. It can be shown that if the graph has N
vertices and we run the walk for $2N^3$ steps, then the probability of not reaching v if
there is a path from u to v is less than one half. So if we run a walk of this length m
times and do not reach v during any of these walks, the probability of this occurring
if there is a path from u to v is less than $1/2^m$. Therefore, we shall say that if during
one of these walks we find v, then there is a path from u to v, and if after m walks
of length $2N^3$ during which we do not reach v, then there is no path from u to v.
Our probability of making a mistake is less than 2^{-m}. This gives us a probabilistic
algorithm for determining whether there is a path from u to v.

There are a number of different ways to define a quantum walk, but we shall
only explore one of them, known as the scattering quantum walk. In this walk, the
particle resides on the edges and can be though of as scattering when it goes through
a vertex. In particular, suppose an edge connects vertices v_1 and v_2. There are two
states corresponding to this edge, and these states are assumed to be orthogonal.
There is the state $|v_1 v_2\rangle$ which corresponds to the particle being on the edge and
going from vertex v_1 to v_2 and the state $|v_2, v_1\rangle$, which corresponds to the particle
being on the edge and going from v_2 to v_1. The set of these states for all of the edges
form an orthonormal basis for the Hilbert space of the walking particle.

Next we need a unitary operator that will advance the walk one time step. We
obtain this operator by combining the action of local unitaries that describe what
happens at the individual vertices. Let us consider a vertex v, and let ω_v be the
linear span of the set of edge states entering v and Ω_v be the span of the set of
edge states leaving v. Because each edge attached to v has two states, one entering
and one leaving v, ω_v and Ω_v have the same dimension. The local unitary, U_v,
at v maps ω_v to Ω_v. We are going to require that the action of U_v be completely
symmetric, that is, we want it to act on all of the edges in the same way. In particular,
suppose there are n edges attached to v. We want the amplitude for the particle to be
reflected back onto the edge from which it entered v to be $-r$ and the amplitude for
it to be transmitted through the vertex and leave by a different edge to be t. That is,
if we denote the vertices attached to v by $1, 2, \ldots n$ and if the particle enters v from

Fig. 7.7 A star graph consists of a central vertex 0 and N outer vertices. The outer vertices are connected to the central vertex by N edges. For the figure $N = 8$

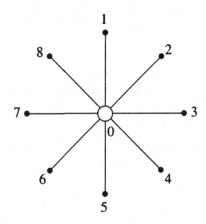

vertex j, then

$$U_v|j,v\rangle = -r|v,j\rangle + t \sum_{k=1,k\neq j}^{n} |v,k\rangle. \qquad (7.45)$$

In order for U_v to be unitary, we must have that the state on the right-hand side of this equation be normalized

$$|r|^2 + (n-1)|t|^2 = 1, \qquad (7.46)$$

and that output states resulting from orthogonal input states be orthogonal

$$-r^*t - rt^* + (n-2)|t|^2 = 0. \qquad (7.47)$$

If, for convenience, we also require that r and t be real, we find that

$$r = \frac{n-2}{n} \qquad t = \frac{2}{n}. \qquad (7.48)$$

Note that with this choice, $r + t = 1$. The action of the unitary operator U that advances walk one step is given by the combined action of all of the operators U_v at the different vertices.

Let us look at a walk on a simple graph known as a star graph, shown in Fig. 7.7. It consists of a central vertex with N edges attached to it and N vertices attached to the other ends of these edges. We shall denote the central vertex by 0 and the outer vertices by $1, 2, \ldots N$. The local unitary corresponding to the central vertex is described by the operator U_v above with $r = (N-2)/N$ and $t = 2/N$. The outer vertices reflect the particle except for one, which we shall assume is vertex 1, that reflects the particle and flips the phase of the state as well. That is the marked vertex, the one that is different from the others, that we are trying to find. Therefore, we have

$U|0,j\rangle = |j,0\rangle$ for $j \geq 2$ and $U|0,1\rangle = -|1,0\rangle$. We shall start the walk in the state

$$|\psi_{init}\rangle = \frac{1}{\sqrt{N}} \sum_{j=1}^{N} |0,j\rangle. \qquad (7.49)$$

Because of the symmetry of the problem the walk takes place in only a subspace of the entire Hilbert space, and the dimension of this subspace is small. In particular, if we define

$$|\psi_1\rangle = |0,1\rangle$$
$$|\psi_2\rangle = |1,0\rangle$$
$$|\psi_3\rangle = \frac{1}{\sqrt{N-1}} \sum_{j=2}^{N} |0,j\rangle$$
$$|\psi_4\rangle = \frac{1}{\sqrt{N-1}} \sum_{j=2}^{N} |j,0\rangle \qquad (7.50)$$

then the action of U on these states is given by

$$U|\psi_1\rangle = -|\psi_2\rangle$$
$$U|\psi_2\rangle = -r|\psi_1\rangle + t\sqrt{N-1}|\psi_3\rangle$$
$$U|\psi_3\rangle = |\psi_4\rangle$$
$$U|\psi_4\rangle = r|\psi_3\rangle + t\sqrt{N-1}|\psi_1\rangle. \qquad (7.51)$$

From this we see that the four-dimensional subspace spanned by these vectors is invariant under U. Our initial state, which can be expressed as

$$|\psi_{init}\rangle = \frac{1}{\sqrt{N}}|\psi_1\rangle + \sqrt{\frac{N-1}{N}}|\psi_3\rangle, \qquad (7.52)$$

is also in this subspace, and so the entire quantum walk will take place in the four-dimensional invariant subspace. This drastically simplifies finding the state of the particle after n steps.

Now from the way this walk has been set up, you might suspect that it will simply mimic the action of the Grover algorithm. If that is the case, you are right. In order to see this, we first note that the action of U in the invariant subspace can be described by a 4×4 matrix

$$M = \begin{pmatrix} 0 & -r & 0 & t\sqrt{N-1} \\ -1 & 0 & 0 & 0 \\ 0 & t\sqrt{N-1} & 0 & r \\ 0 & 0 & 1 & 0 \end{pmatrix}, \qquad (7.53)$$

where the matrix elements of M are given by $M_{jk} = \langle \psi_j | U | \psi_k \rangle$. In order to find out how the walk behaves, we first find the eigenvalues and eigenvectors of U. The characteristic equation for the eigenvalues, λ, of M is

$$\lambda^4 - 2r\lambda^2 + 1 = 0. \tag{7.54}$$

We will solve the equation in the large N limit. In that case, we express the equation as

$$\lambda^4 - 2\lambda^2 + 1 + 2t\lambda^2 = 0. \tag{7.55}$$

We ignore the last term on the left-hand side, which is small when N is large, in order to find zeroth order solutions, λ_0. This gives $\lambda_0 = \pm 1$. We now set $\lambda = \lambda_0 + \delta\lambda$, and substitute it back into the equation. Keeping terms of up to second order in small quantities we find that for $\lambda_0 = 1$

$$\delta\lambda^2 + \frac{1}{2}t(1 + 2\delta\lambda) = 0, \tag{7.56}$$

and for $\lambda_0 = -1$ we find

$$\delta\lambda^2 + \frac{1}{2}t(1 - 2\delta\lambda) = 0. \tag{7.57}$$

In both cases, the solutions are to lowest order in $1/N$

$$\delta\lambda = \pm i\sqrt{\frac{t}{2}}, \tag{7.58}$$

which is of order $N^{-1/2}$.

It is also necessary to find the eigenstates of M. Setting $\Delta = \sqrt{t/2}$, we have that for $\lambda = 1 + i\Delta$ and $\lambda = 1 - i\Delta$, the eigenstates are, respectively,

$$|u_1\rangle = \frac{1}{2}\begin{pmatrix} -1 \\ 1 \\ -i \\ -i \end{pmatrix} \qquad |u_2\rangle = \frac{1}{2}\begin{pmatrix} -1 \\ 1 \\ i \\ i \end{pmatrix}, \tag{7.59}$$

and for $\lambda = -1 + i\Delta$ and $\lambda = -1 - i\Delta$, the eigenstates are, respectively,

$$|u_3\rangle = \frac{1}{2}\begin{pmatrix} 1 \\ 1 \\ -i \\ i \end{pmatrix} \qquad |u_4\rangle = \frac{1}{2}\begin{pmatrix} 1 \\ 1 \\ i \\ -i \end{pmatrix}. \tag{7.60}$$

In terms of the eigenstates, we see that

$$|\psi_{init}\rangle = \frac{i}{2}(|u_1\rangle - |u_2\rangle + |u_3\rangle - |u_4\rangle) + O(N^{-1/2}). \qquad (7.61)$$

Noting that $1 \pm i\Delta \cong e^{\pm i\Delta}$ and $-1 \pm i\Delta \cong -e^{\mp i\Delta}$, we have that

$$U^n|\psi_{init}\rangle = \frac{i}{2}[e^{in\Delta}|u_1\rangle - e^{-in\Delta}|u_2\rangle$$
$$+ (-1)^n(e^{-in\Delta}|u_3\rangle - e^{in\Delta}|u_4\rangle)] + O(N^{-1/2}), \qquad (7.62)$$

or

$$U^n|\psi_{init}\rangle = \frac{1}{2}\begin{pmatrix} \sin(n\Delta) \\ -\sin(n\Delta) \\ \cos(n\Delta) \\ \cos(n\Delta) \end{pmatrix} + \frac{1}{2}(-1)^n\begin{pmatrix} \sin(n\Delta) \\ \sin(n\Delta) \\ \cos(n\Delta) \\ -\cos(n\Delta) \end{pmatrix}, \qquad (7.63)$$

up to order $N^{-1/2}$.

From this result, we see that when $n\Delta$ is close to $\pi/2$, the particle will be located on the edge connected to the marked vertex. If n is even it will be in the state $|0,1\rangle$ and if n is odd it will be in the state $-|1,0\rangle$. By simply measuring the location of the particle, in particular, which edge it is on, we will find which vertex is the marked one. Note that if $n\Delta$ is close to $\pi/2$, then n is of order $\sqrt{N}$. Classically, in order to find the marked vertex, we would have to check each vertex, which would require $O(N)$ operations, whereas if we run a quantum walk, we can find the marked vertex in $O(\sqrt{N})$ steps. Therefore, we obtain a quadratic speedup.

So far, we have only used a quantum walk to do something we already knew how to do, find a marked element in a list. Let us see if we can use it to do something else. Suppose that instead of a marked vertex, our star graph has an extra edge. That is, there is an edge between two of the outer vertices, and we would like to find out where it is. A quantum walk can provide a quadratic speedup for this type of search as well. The graph is depicted in Fig. 7.8.

Let's assume the extra edge is between vertices 1 and 2. That means that besides the states $|0,j\rangle$ and $|j,0\rangle$, for $j = 1,2,\ldots N$, we also have the states $|1,2\rangle$ and $|2,1\rangle$. For simplicity we shall assume that vertices 1 and 2 just transmit the particle. Our unitary operator will now act as $U|0,j\rangle = |j,0\rangle$ for $j > 2$, and

$$U|0,1\rangle = |1,2\rangle \quad U|0,2\rangle = |2,1\rangle$$
$$U|1,2\rangle = |2,0\rangle \quad U|2,1\rangle = |1,0\rangle. \qquad (7.64)$$

Its action on the states $|j,0\rangle$ is as before. The walk resulting from this choice of U can also be analyzed easily, because it stays within a five-dimensional subspace of the entire Hilbert space. Define the states

Fig. 7.8 A star graph with an extra edge between two outer vertices

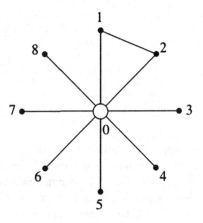

$$|\psi_1\rangle = \frac{1}{\sqrt{2}}(|0,1\rangle + 0,2\rangle)$$

$$|\psi_2\rangle = \frac{1}{\sqrt{2}}(|1,0\rangle + |2,0\rangle)$$

$$|\psi_3\rangle = \frac{1}{\sqrt{N-2}}\sum_{j=3}^{N}|0,j\rangle$$

$$|\psi_4\rangle = \frac{1}{\sqrt{N-2}}\sum_{j=3}^{N}|j,0\rangle$$

$$|\psi_5\rangle = \frac{1}{\sqrt{2}}(|1,2\rangle + |2,1\rangle). \tag{7.65}$$

These states span a five-dimensional space we shall call S. The unitary transformation, U, that advances the walk one step acts on these states as follows:

$$U|\psi_1\rangle = |\psi_5\rangle$$

$$U|\psi_2\rangle = -(r-t)|\psi_1\rangle + 2\sqrt{rt}|\psi_3\rangle$$

$$U|\psi_3\rangle = |\psi_4\rangle$$

$$U|\psi_4\rangle = (r-t)|\psi_3\rangle + 2\sqrt{rt}|\psi_1\rangle$$

$$U|\psi_5\rangle = |\psi_2\rangle. \tag{7.66}$$

For our initial state we choose

$$|\psi_{init}\rangle = \frac{1}{\sqrt{2N}} \sum_{j=1}^{N} (|0,j\rangle - |j,0\rangle)$$

$$= \frac{1}{\sqrt{N}} (|\psi_1\rangle - |\psi_2\rangle)$$

$$+ \sqrt{\frac{N-2}{2N}} (|\psi_3\rangle - |\psi_4\rangle), \tag{7.67}$$

which is in S. Since the initial state is in S and S is an invariant subspace of U, the entire walk will remain in S, and so we find ourselves in a situation similar to the previous one. This search, however, is more sensitive to the choice of initial state than the previous one. While we didn't mention it before, in the previous search we could also have taken a superposition of all ingoing states instead of all outgoing ones as our initial state. In the present case, the minus sign in the first expression for initial state is essential; if it is replaced by a plus sign, the search will fail.

In order to find the evolution of the quantum state for this walk, we proceed as before and find the eigenvalues and eigenstates of U restricted to S. The matrix that describes the action of U on S is given by

$$M = \begin{pmatrix} 0 & -(r-t) & 0 & 2\sqrt{rt} & 0 \\ 0 & 0 & 0 & 0 & 1 \\ 0 & 2\sqrt{rt} & 0 & (r-t) & 0 \\ 0 & 0 & 1 & 0 & 0 \\ 1 & 0 & 0 & 0 & 0 \end{pmatrix}. \tag{7.68}$$

The characteristic equation for this matrix is

$$\lambda^5 - (r-t)\lambda^3 + (r-t)\lambda^2 - 1 = 0. \tag{7.69}$$

One root of this equation is $\lambda = 1$, and if we factor out $(\lambda - 1)$ from the above equation, we are left with

$$\lambda^4 + \lambda^3 + 2t\lambda^2 + \lambda + 1 = 0. \tag{7.70}$$

As before, we will use a perturbation expansion to find the roots of this equation with the transmission amplitude, t, as the small parameter. The zeroth order solutions are found by setting $t = 0$, which gives us the large N limit, and we find

$$\lambda^4 + \lambda^3 + \lambda + 1 = (\lambda + 1)(\lambda^3 + 1) = 0, \tag{7.71}$$

so the zeroth order roots are -1 twice, $e^{i\pi/3}$, and $e^{-i\pi/3}$. Setting $\lambda = -1 + \delta\lambda$, substituting into the above equation and keeping terms of up to $(\delta\lambda)^2$ gives

$$3(\delta\lambda)^2 - 4t\delta\lambda + 2t = 0, \tag{7.72}$$

whose solution, keeping lowest order terms, is

$$\delta\lambda = \pm i\sqrt{\frac{2t}{3}} = O(N^{-1/2}).\tag{7.73}$$

If we set $\lambda = e^{\pm i\pi/3} + \delta\lambda$, we find that $\delta\lambda = O(N^{-1})$, so these roots and their corresponding eigenvalues are not of interest, because they will not yield a quadratic speedup.

We now need to find the eigenvectors. If the components of the eigenvectors are denoted by x_j, where $j = 1, \ldots 5$, the eigenvector equations are

$$- (r-t)x_2 + 2\sqrt{rt}x_4 = (-1 \pm i\Delta)x_1$$
$$x_5 = (-1 \pm i\Delta)x_2$$
$$2\sqrt{rt}x_2 + (r-t)x_4 = (-1 \pm i\Delta)x_3$$
$$x_3 = (-1 \pm i\Delta)x_4$$
$$x_1 = (-1 \pm i\Delta)x_5,\tag{7.74}$$

where now $\Delta = (2t/3)^{1/2}$. To lowest order (the terms that were dropped are of order $1/\sqrt{N}$ or smaller) the eigenvector corresponding to the eigenvalue $-1 + i\Delta$ is

$$|v_1\rangle = \frac{1}{\sqrt{6}}\begin{pmatrix} 1 \\ 1 \\ -i\sqrt{3/2} \\ i\sqrt{3/2} \\ -1 \end{pmatrix},\tag{7.75}$$

and the eigenvector corresponding to the eigenvalue $-1 - i\Delta$ is

$$|v_2\rangle = \frac{1}{\sqrt{6}}\begin{pmatrix} 1 \\ 1 \\ i\sqrt{3/2} \\ -i\sqrt{3/2} \\ -1 \end{pmatrix}.\tag{7.76}$$

We find that, up to terms of order $N^{-1/2}$, our initial state can be expressed as

$$|\psi_{init}\rangle = \frac{i}{\sqrt{2}}(|v_1\rangle - |v_2\rangle).\tag{7.77}$$

Expressing the eigenvalues corresponding to $|v_1\rangle$ and $|v_2\rangle$ as

$$-1 + i\Delta \cong -e^{-i\Delta} \qquad -1 - i\Delta \cong -e^{i\Delta}\tag{7.78}$$

we find that the state after n steps is

$$U^n|\psi_{init}\rangle = \frac{(-1)^n}{\sqrt{3}}\begin{pmatrix} \sin(n\Delta) \\ \sin(n\Delta) \\ \sqrt{3/2}\cos(n\Delta) \\ -\sqrt{3/2}\cos(n\Delta) \\ -\sin(n\Delta) \end{pmatrix}. \tag{7.79}$$

From this equation, we can see that when $n\Delta = \pi/2$, the particle is located on one of the edges leading to the extra edge or on the extra edge itself. This will happen when $n = O(\sqrt{N})$.

We now need to discuss how to interpret this result. It is reasonable to assume that if we are given a graph with an extra edge in an unknown location, we only have access to the edges connecting the central vertex to the outer ones, and not to the extra edge itself (if we had access to the extra edge, then we would have to know where it is). That is, in making a measurement, we can only determine which of the edges connecting central vertex to the outer ones the particle is on. If it is on the extra edge, we will not detect it. So, after n steps, where $n\Delta = \pi/2$, we measure the edges to which we have access to find out where the particle is. With probability $2/3$ it will be on an edge connected to the extra edge, and with probability $1/3$, it will be on the extra edge itself, in which case we won't detect it.

In comparing this procedure to a classical search for the extra edge, we shall assume that classically the graph is specified by an adjacency list, which is an efficient specification for sparse graphs. For each vertex of the graph, one lists the vertices that are connected to it by an edge. In our case, the central vertex is connected to all of the other vertices, the vertices not connected to the extra edge are connected only to the central vertex, and two of the outer vertices are connected to the central vertex and to each other. Searching this list classically would require $O(N)$ steps to find the extra edge, while the quantum procedure will succeed in $O(\sqrt{N})$ steps. Therefore, we again obtain a quadratic speedup by using a quantum walk.

We have just examined the use of quantum walks in search problems, but they have been useful in developing other types of algorithms as well. One example is element distinctness. One has a function in the form of a black box, that is, one puts in an input x and the output is $f(x)$, but we have no knowledge about the function. We can only send in inputs and obtain outputs. Our task is to find two inputs, if they exist, that give the same output. This can be accomplished by using a kind of quantum walk, which requires fewer queries to the black box than is necessary on a classical computer. It is also possible to use quantum walks to evaluate certain types of Boolean formulas with fewer queries than are possible classically.

7.7 Problems

1. Suppose we have a star graph with a loop on one of its outer vertices, say vertex 1. The other vertices simply reflect the particle, $U|0,j\rangle = |j,0\rangle$, for $j > 1$. The loop has one quantum state, which we shall denote by $|l_1\rangle$. The unitary operator has the action $U|0,1\rangle = |l_1\rangle$ and $U|l_1\rangle = |1,0\rangle$. Show that starting with the initial state in Eq. (7.67) the particle making the walk will become localized on the loop and the edge connected to the loop in $O(\sqrt{N})$ steps.

2. We have a Controlled-U gate, which acts on two qubits. Qubit a is the control qubit and qubit b is the target qubit, so that $|0\rangle_a|j\rangle_b \rightarrow |0\rangle_a|j\rangle_b$ and $|1\rangle_a|j\rangle_b \rightarrow |1\rangle_a U|j\rangle_b$ for $j = 0,1$. Suppose that the eigenvalues of U are ± 1. Our task is to generate two qubits, one in the $+1$ eigenstate, $|u_+\rangle$, and one in the -1 eigenstate, $|u_-\rangle$, with one use of the gate. Show, making use of the rotational invariance of the singlet state

$$|\phi_s\rangle = \frac{1}{\sqrt{2}}(|0\rangle|1\rangle - |1\rangle|0\rangle),$$

that if we start with the three-qubit state

$$|\Psi_{in}\rangle_{abc} = |+x\rangle_a|\phi_s\rangle_{bc}$$

and send qubits a and b through the Controlled-U gate and make the proper measurement of qubit a, then one of the remaining qubits will be in the state $|u_+\rangle$ and the other will be in the state $|u_-\rangle$, and we will know which qubit is in which state.

3. Suppose we have a black box that evaluates the Boolean function $f(x)$, where x is the n-bit string $x_1, x_2, \ldots x_n$. This function is a sum of linear and quadratic terms in the variables x_j and each variable appears in only one term. By considering the function $f(x) + f(\bar{x})$, where $\bar{x}$ is the n-bit string $x_1 + 1, x_2 + 1, \ldots x_n + 1$, show that we can use the Bernstein–Vazirani algorithm to determine which variables appear in quadratic terms with two function evaluations. How many evaluations would be required classically?

References

1. R. Cleve, A. Ekert, C. Macchiavello, M. Mosca, Quantum algorithms revisited. Proc. R. Soc. Lond. A **454**, 339 (1998)
2. E. Bernstein, U. Vazirani, Quantum complexity theory. SIAM J. Comput. **26**, 1411 (1997)
3. L.K. Grover, Quantum mechanics helps in searching for a needle in a haystack. Phys. Rev. Lett. **79**, 325 (1997)

4. D. Aharonov, A. Ambainis, J. Kempe, U. Vazirani, Quantum walks on graphs. In *Proceedings of the 33rd Symposium on the Theory of Computing (STOC01)* (ACM Press, New York, 2001), pp. 50–59 and quant-ph/0012090
5. E. Feldman, M. Hillery, H.-W. Lee, D. Reitzner, H. Zheng, V. Bužek, Finding structural anomalies in graphs by means of quantum walks. Phys. Rev. A **82**, 040302R (2010)
6. A.M. Childs, W. van Dam, Quantum algorithms for algebraic problems. Rev. Mod. Phys. **82**, 1 (2010)

Chapter 8
Quantum Machines

8.1 Introduction

If we are given quantum information in the form of qubits in a particular state, we have seen that we can process that information by sending the qubits through different sequences of gates. A particular collection of gates constitutes a quantum machine that manipulates the information encoded in the qubits in a particular way. A quantum machine can perform either a single task or, if it is programmable, a number of different tasks, the exact task depending on the program.

In this chapter we want to look at several different quantum machines. The first two, a quantum cloner and a Universal-NOT (or, more concisely, U-NOT) gate, are single task machines. They both perform, approximately, tasks that cannot be performed exactly. We then move on to programmable machines. We will first prove a general result that shows there is no deterministic, universal programmable quantum processor. We will then examine two different probabilistic programmable machines. The first is based on the same circuit as the cloner. We will show, as an example, how it can be used to implement a quantum phase gate in any basis, the basis being determined by the program. The second unambiguously discriminates between two states, but the two states it is discriminating between are given in the form of a program and not hardwired into the machine.

8.2 Cloners and U-NOT Gates

As we have seen, a device that perfectly clones a quantum state is impossible to construct. However, if we relax the requirement that the copies be perfect, it is possible to copy quantum information. A second operation that is not possible to perform exactly is the U-NOT operation. This ideally would take a qubit in an arbitrary state $|\psi\rangle = \alpha|0\rangle + \beta|1\rangle$ and send it into the orthogonal state $|\psi_{\perp}\rangle = \beta^{*}|0\rangle - \alpha^{*}|1\rangle$. The indication that this is an impossible operation is the appearance

J.A. Bergou and M. Hillery, *Introduction to the Theory of Quantum Information Processing*, Graduate Texts in Physics, DOI 10.1007/978-1-4614-7092-2_8, © Springer Science+Business Media New York 2013

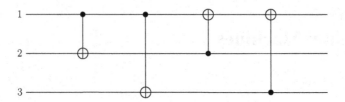

Fig. 8.1 Quantum circuit for the cloning machine

of complex conjugates. The perfect U-NOT operation is an anti-unitary one, but quantum operations must be unitary. It is, nonetheless, possible to construct an approximate U-NOT gate, and, it turns out, approximate cloners and approximate U-NOT gates are closely related.

Let us begin with the cloner. Consider the circuit shown in Fig. 8.1.

The circuit consists of three qubits being acted upon by four Controlled-NOT gates. The input qubit is qubit number 1, and it is its state we wish to copy. In order to see how this works, let us consider what happens with different input states for the remaining two qubits. Define the two two-qubit states

$$|\Xi_{00}\rangle = \frac{1}{\sqrt{2}}(|0\rangle|0\rangle + |1\rangle|1\rangle)$$

$$|\Xi_{0x}\rangle = \frac{1}{\sqrt{2}}|0\rangle(|0\rangle + |1\rangle). \tag{8.1}$$

Now if qubit 1 is in the state $|\psi\rangle_1$ and qubits 2 and 3 are in one of the two states above, then the cloning circuit will implement the following transformations:

$$|\psi\rangle_1|\Xi_{00}\rangle_{23} \rightarrow |\psi\rangle_1|\Xi_{00}\rangle_{23}$$

$$|\psi\rangle_1|\Xi_{0x}\rangle_{23} \rightarrow |\psi\rangle_2|\Xi_{00}\rangle_{13}. \tag{8.2}$$

Examining these equations, we see that in the first the quantum information from the first qubit appears in output 1, and in the second it appears in output 2, so what this circuit does is move the information from the first qubit around, and the location to which it gets moved is determined by the state sent into inputs 2 and 3. This suggests that if instead of sending either $|\Xi_{00}\rangle$ or $|\Xi_{0x}\rangle$ into inputs 2 and 3, we send in a linear combination of them, some of the quantum information from qubit 1 will appear in output 1 and some of it will appear in output 2, thereby cloning the state. This is, in fact, exactly what happens. If we choose

$$|\Psi\rangle_{23} = c_0|\Xi_{00}\rangle_{23} + c_1|\Xi_{0x}\rangle_{23}, \tag{8.3}$$

as the input state for qubits 2 and 3, with c_0 and c_1 real for simplicity, and $c_0^2 + c_1^2 + c_0 c_1 = 1$ so that the state is normalized, the reduced density matrices for outputs 1 and 2 are

$$\rho_1^{(\text{out})} = (c_0^2 + c_0 c_1)|\psi\rangle\langle\psi| + \frac{c_1^2}{2}I$$

$$\rho_2^{(\text{out})} = (c_1^2 + c_0 c_1)|\psi\rangle\langle\psi| + \frac{c_0^2}{2}I. \tag{8.4}$$

Note that by choosing c_0 and c_1 we can control how much information about $|\psi\rangle$ goes to which output. In particular, if we choose $c_0 = c_1 = 1/\sqrt{3}$, then the information is divided equally, and we find that

$$\rho_1^{(\text{out})} = \rho_2^{(\text{out})} = \frac{5}{6}|\psi\rangle\langle\psi| + \frac{1}{6}|\psi_\perp\rangle\langle\psi_\perp|, \tag{8.5}$$

where $|\psi_\perp\rangle$ is the qubit state orthogonal to $|\psi\rangle$. Therefore, the fidelity of the cloner output $\rho_1^{(\text{out})}$ (or $\rho_2^{(\text{out})}$ since they are the same in this case) to the ideal output, $|\psi\rangle$, which is given by $\langle\psi|\rho_1^{(\text{out})}|\psi\rangle$, is 5/6. A fidelity of one would imply perfect cloning, so what we have here is a device that produces two copies of the input qubit that are pretty good approximations to it. Note that the fidelity does not depend on the input state, that is, all states are cloned equally well. This feature of this cloning machine is known as universality.

Note that the cloner employs three qubits, and we have only discussed the final state of two of them. One might wonder if the output state of the third qubit is of interest. This is, in fact, where the connection with the U-NOT gate enters. The output state of the third qubit is given by

$$\rho_3^{(\text{out})} = c_0 c_1|\psi^*\rangle\langle\psi^*| + \frac{1}{2}(1 - c_0 c_1)I, \tag{8.6}$$

where $|\psi^*\rangle = \alpha^*|0\rangle + \beta^*|1\rangle$ and I is the two-by-two identity matrix. If we now apply the unitary operator $U_0 = -i\sigma_y$, which has the effect $U_0|0\rangle = -|1\rangle$ and $U_0|1\rangle = |0\rangle$, to this density matrix, and make use of the fact that $I = |\psi\rangle\langle\psi| + |\psi_\perp\rangle\langle\psi_\perp|$, we find

$$U_0\rho_3^{(\text{out})}U_0^{-1} = \frac{1}{2}(1 + c_0 c_1)|\psi_\perp\rangle\langle\psi_\perp| + \frac{1}{2}(1 - c_0 c_1)|\psi\rangle\langle\psi|. \tag{8.7}$$

In the case that $c_0 = c_1 = 1/\sqrt{3}$ this becomes

$$U_0\rho_3^{(\text{out})}U_0^{-1} = \frac{2}{3}|\psi_\perp\rangle\langle\psi_\perp| + \frac{1}{3}|\psi\rangle\langle\psi|. \tag{8.8}$$

This is, in fact, the best approximation to the state orthogonal to that of the input qubit that can be realized, a fact we will not prove here. Note that the fidelity of the output to the ideal output state, $|\psi_\perp\rangle$, is 2/3. Therefore, the cloner with the addition of a U_0 gate to the third output not only clones states, but it also realizes the best possible approximate U-NOT gate.

The same result for the U-NOT operation can be achieved by measuring the original qubit. We measure $|\psi\rangle$ along a random direction in our two-dimensional Hilbert space

$$|\eta\rangle = \cos(\theta/2)|0\rangle + e^{i\phi}\sin(\theta/2)|1\rangle, \tag{8.9}$$

that is, we measure the projection $|\eta\rangle\langle\eta|$. If we obtain the result 1, we produce the state $|\eta_\perp\rangle$, where

$$|\eta_\perp\rangle = e^{-i\phi}\sin(\theta/2)|0\rangle - \cos(\theta/2)|1\rangle, \tag{8.10}$$

and if we get 0, we produce the state $|\eta\rangle$. The density matrix resulting from this procedure is

$$\rho^{(\text{out})}(\eta) = |\langle\psi|\eta\rangle|^2|\eta_\perp\rangle\langle\eta_\perp| + |\langle\psi|\eta_\perp\rangle|^2|\eta\rangle\langle\eta|. \tag{8.11}$$

If we now average this over η we find

$$\begin{aligned}
\rho^{(\text{out})} &= \frac{1}{4\pi}\int_0^{2\pi}\mathrm{d}\phi\int_0^{\pi}\mathrm{d}\theta\sin(\theta)\rho^{(\text{out})}(\eta) \\
&= \frac{2}{3}|\psi_\perp\rangle\langle\psi_\perp| + \frac{1}{3}|\psi\rangle\langle\psi|.
\end{aligned} \tag{8.12}$$

Therefore, the best approximate U-NOT gate can be achieved in two different ways: One is to use the cloning circuit, and the second is to the measure the qubit in a random direction and then produce a qubit whose direction is opposite to that indicated by our measurement result.

One might wonder if a similar strategy can be applied to cloning. That is, one measures the original qubit in a random direction $|\eta\rangle$, and if one gets 1, one produces two qubits in the state $|\eta\rangle|\eta\rangle$, and if one gets 0, one produces two qubits in the state $|\eta_\perp\rangle|\eta_\perp\rangle$. This procedure does work, but, unlike in the case of the U-NOT, it is not optimal. One finds after averaging over η that the fidelity of the output state to the ideal output state, $|\psi\rangle|\psi\rangle$, is $2/3$, which is less than the $5/6$ achieved by the cloning circuit.

8.3 Programmable Machines: A General Result

We now want to consider programmable quantum machines, which we shall often refer to as quantum processors. Programmable machines have a number of advantages over machines that perform a single function. First, they are much more flexible. In order to change what they do, you just change the program rather than rewiring the entire quantum circuit. Second, they offer the possibility of performing several operations on the data in parallel, by using superpositions of program states, where each element of the superposition corresponds to a different operation.

Programmable machines have two inputs, one for the data, which is to be acted upon, and one for the program, which will specify the operation to be performed on the data. Both the data and the program are quantum states. In particular, the processor is a unitary operator acting on the Hilbert space $\mathcal{H}_d \otimes \mathcal{H}_p$, where $\mathcal{H}_d$ is the data Hilbert space and $\mathcal{H}_p$ is the program Hilbert space. Ideally, we would like to be able to program any unitary operator acting on the data. For example, if our data space is two-dimensional, we would like to be able to have a program for each element of $SU(2)$. Such a processor would be universal, that is, it could be used to deterministically perform any unitary operation on a qubit. Unfortunately, as shown by Nielsen and Chuang, it is impossible to construct such a processor.

In order to show this, we need to examine the resources that are necessary in order to implement a given set of operations on the data. What Nielsen and Chuang showed is that if the program $|\Xi_1\rangle_p \in \mathcal{H}_p$ implements the unitary operator U_1 on the data state, and $|\Xi_2\rangle_p \in \mathcal{H}_p$ implements the unitary operator U_2, then $_p\langle\Xi_1|\Xi_2\rangle_p = 0$. This implies that for every unitary operator that the processor can implement on the data state, we need an extra dimension in the program space. Since the number of operations in $SU(2)$ is uncountably infinite, a program space that is finite, or even countably finite, would not be big enough to account for every operation.

Let us now prove the no-go theorem for deterministic programmable quantum processors. We assume the processor is represented by a unitary operator, G, acting in $\mathcal{H}_d \otimes \mathcal{H}_p$, where $\mathcal{H}_d$ is the data space and $\mathcal{H}_p$ is the program space. We suppose that we have a program $|\Xi_1\rangle_p \in \mathcal{H}_p$ that implements the unitary operator U_1 on $\mathcal{H}_d$, in particular

$$G(|\psi\rangle_d \otimes |\Xi_1\rangle_p) = U_1|\psi\rangle_d \otimes |\Xi_1'\rangle_p. \tag{8.13}$$

Now it could be the case that the output in the program space depends on the state $|\psi\rangle_d$ that is sent into the data input. In order to show that this is not the case, assume that

$$G(|\psi_1\rangle_d \otimes |\Xi_1\rangle_p) = U_1|\psi_1\rangle_d \otimes |\Xi_1'\rangle_p$$
$$G(|\psi_2\rangle_d \otimes |\Xi_1\rangle_p) = U_1|\psi_2\rangle_d \otimes |\Xi_1''\rangle_p. \tag{8.14}$$

Taking the inner products of the left-hand sides of the above equations and equating that to the inner product of the right-hand sides, and assuming that $_d\langle\psi_1|\psi_2\rangle_d \neq 0$, give us $_p\langle\Xi_1'|\Xi_1''\rangle_p = 1$, thereby implying that the program state outputs are identical.

Now suppose that the program state $|\Xi_1\rangle_p$ implements the operator U_1 and the program state $|\Xi_2\rangle_p$ implements U_2. We then have that

$$G(|\psi\rangle_d \otimes |\Xi_1\rangle_p) = U_1|\psi\rangle_d \otimes |\Xi_1'\rangle_p$$
$$G(|\psi\rangle_d \otimes |\Xi_2\rangle_p) = U_2|\psi\rangle_d \otimes |\Xi_1'\rangle_p. \tag{8.15}$$

Taking inner products we find

$$_p\langle\Xi_2|\Xi_1\rangle_p = {}_d\langle\psi|U_2^{-1}U_1|\psi\rangle_d \, {}_p\langle\Xi_2'|\Xi_1'\rangle_p. \tag{8.16}$$

We will examine both the case $_p\langle\Xi_2'|\Xi_1'\rangle_p \neq 0$ and the case $_p\langle\Xi_2'|\Xi_1'\rangle_p = 0$. If $_p\langle\Xi_2'|\Xi_1'\rangle_p \neq 0$, we have

$$\frac{_p\langle\Xi_2|\Xi_1\rangle_p}{_p\langle\Xi_2'|\Xi_1'\rangle_p} = {}_d\langle\psi|U_2^{-1}U_1|\psi\rangle_d, \tag{8.17}$$

and we note that the left-hand side does not depend on $|\psi\rangle_d$, so the right-hand side cannot either. That implies that $U_2^{-1}U_1$ is a multiple of the identity, and since both of the operators are unitary, we must have $U_2 = e^{i\theta}U_1$ for some θ between 0 and 2π. Now if, on the other hand, $_p\langle\Xi_2'|\Xi_1'\rangle_p = 0$, then we see that we must also have that $_p\langle\Xi_2|\Xi_1\rangle_p = 0$. Summarizing, what we have found is that if U_1 and U_2 are different, that is, they are not multiples of each other, then they must correspond to orthogonal program states. Therefore, the dimension of the program space must be at least as great as the number of unitary operators that the processor can perform.

Similar reasoning can be employed to show that a deterministic scheme employing measurement is also impossible. We can call this a measure-and-correct scheme. Suppose that we send a program and data into our processor, and at the output measure the program state in a fixed basis. Each measurement outcome corresponds to a different unitary operator being applied to the data state, but for each program state the resulting operators are related to each other in the same way. That means that for any program state, if we do not obtain the desired measurement result, we can correct the resulting output state by applying an operator that does not depend on the program state.

Let us look at a simple example. Suppose that both the data and program spaces are two-dimensional and that our processor acts as follows:

$$G(|\psi\rangle_d \otimes |\Xi_1\rangle_p) = \frac{1}{\sqrt{2}}(U_1|\psi\rangle_d \otimes |0\rangle_p + VU_1|\psi\rangle_d \otimes |1\rangle_p)$$

$$G(|\psi\rangle_d \otimes |\Xi_2\rangle_p) = \frac{1}{\sqrt{2}}(U_2|\psi\rangle_d \otimes |0\rangle_p + VU_2|\psi\rangle_d \otimes |1\rangle_p). \tag{8.18}$$

Here, V is a fixed unitary operator. Such a processor is capable of deterministically applying four different unitary operators to the data state, U_1, VU_1, U_2, and VU_2. For example, suppose we want to apply U_1. We use the program $|\Xi_1\rangle$ and then measure the program state in the basis $\{|0\rangle, |1\rangle\}$. If we obtain $|0\rangle$ we are done, and if we obtain $|1\rangle$, then we can apply V^{-1} to the data state. In either case, we obtain the output state $U_1|\psi\rangle_d$. We will also be able to deterministically obtain the superpositions $c_1U_1 + c_2U_2$ and $c_1VU_1 + c_2VU_2$, where c_1 and c_2 are complex numbers. It appears that we have beaten the no-go theorem, because we are able to deterministically realize four unitary operators with a two-dimensional program space. Unfortunately, it will not work. If we take the inner products of the two equations above, we find that

$$_p\langle\Xi_1|\Xi_2\rangle_p = {}_d\langle\psi|U_1^{-1}U_2|\psi\rangle_d. \tag{8.19}$$

The left-hand side does not depend on $|\psi\rangle_d$, which, as before, implies that U_1 and U_2 are related by a phase factor and that the program states are multiples of each other. Therefore, we can only realize two operators in this way, U_1 and VU_1, and we have not gained anything.

8.4 Probabilistic Processors

The no-go result we proved in the previous section only applies to deterministic processors. If the processor is probabilistic, its limitations no longer apply. Let us first illustrate this with a simple example and then proceed to a more complicated one. Suppose our data system is a qubit, and we want to implement the one parameter group of transformations, $U(\alpha) = \exp(i\alpha\sigma_z)$, where $0 \leq \alpha < 2\pi$. This can be accomplished with a success probability of $1/2$ by using a qubit program and a Controlled-NOT gate. The Controlled-NOT gate has two inputs, a control input and a target input, and in the case we wish to consider here, the target qubit is the program and the control qubit is the data. The program states are

$$|\Xi(\alpha)\rangle = \frac{1}{\sqrt{2}}(e^{i\alpha}|0\rangle + e^{-i\alpha}|1\rangle). \tag{8.20}$$

If the data state input is $|\psi\rangle$, the output of this processor is then

$$|\Psi_{\text{out}}\rangle = \frac{1}{\sqrt{2}}(U(\alpha)|\psi\rangle|0\rangle + U^{-1}(\alpha)|\psi\rangle|1\rangle). \tag{8.21}$$

By measuring the program state output in the basis $\{|0\rangle, |1\rangle\}$ and keeping the result only if we get $|0\rangle$, which happens with a probability of $1/2$, we obtain the data state output $U(\alpha)|\psi\rangle$, which is the desired result. Note that in this case, a single processor is able to realize, with a one-qubit program space, a continuous group of transformations. The cost is that in each application, the desired transformation is only realized with a probability of $1/2$.

Now let us look at a more complicated example. We will begin by going back and considering the three-qubit circuit for the approximate cloner. Qubit 1 will now be our data state, and qubits 2 and 3 will be our program. We will denote the data state by $|\psi\rangle_1$ and the program state by $|\Xi\rangle_{23}$. Define the two-qubit Bell states to be

$$|\Psi_{\pm}\rangle = \frac{1}{\sqrt{2}}(|00\rangle \pm |11\rangle)$$

$$|\Phi_{\pm}\rangle = \frac{1}{\sqrt{2}}(|01\rangle \pm |10\rangle). \tag{8.22}$$

If these states are used as programs in our processor, we find that

$$|\psi\rangle_1|\Psi_+\rangle_{23} \rightarrow |\psi\rangle_1|\Psi_+\rangle_{23}$$

$$|\psi\rangle_1|\Psi_-\rangle_{23} \rightarrow \sigma_z|\psi\rangle_1|\Psi_-\rangle_{23}$$

$$|\psi\rangle_1|\Phi_+\rangle_{23} \rightarrow \sigma_x|\psi\rangle_1|\Phi_+\rangle_{23}$$

$$|\psi\rangle_1|\Phi_-\rangle_{23} \rightarrow (-i\sigma_y)|\psi\rangle_1|\Phi_-\rangle_{23}, \tag{8.23}$$

where σ_x, σ_y, and σ_z are the Pauli matrices. Suppose we want to implement the operator

$$U_\phi = |\phi_\perp\rangle\langle\phi_\perp| - |\phi\rangle\langle\phi| = I - 2|\phi\rangle\langle\phi| \tag{8.24}$$

on the data state, where $|\phi\rangle$ and $|\phi_\perp\rangle$ are specified, orthogonal one-qubit states. The operator U_ϕ is similar to σ_z, but instead of flipping the phase of the state $|1\rangle$ and leaving $|0\rangle$ unchanged, it flips the phase of the state $|\phi\rangle$ and leaves the phase of $|\phi_\perp\rangle$ unchanged. In order to find a program state that will implement this operator, we first express it in terms of the Pauli matrices. Setting $|\phi\rangle = \mu|0\rangle + v|1\rangle$, we find

$$U_\phi = -(\mu v^* + \mu^* v)\sigma_x + (\mu v^* - \mu^* v)(-i\sigma_y)$$

$$+ (|v|^2 - |\mu|^2)\sigma_z. \tag{8.25}$$

We can now apply the operation U_ϕ to $|\psi\rangle_1$ by sending in the program state

$$|\Xi_\phi\rangle = -(\mu v^* + \mu^* v)|\Phi_+\rangle_{23} + (\mu v^* - \mu^* v)|\Phi_-\rangle_{23}$$

$$+ (|v|^2 - |\mu|^2)|\Psi_-\rangle_{23}, \tag{8.26}$$

and measuring the program outputs to see if they are in the state $(|\Phi_+\rangle_{23} + |\Phi_-\rangle_{23} + |\Phi_-\rangle_{23})/\sqrt{3}$. This will occur with a probability of $1/3$. When we do obtain this result, the output of the data state is $U_\phi|\psi\rangle_1$. Note that both the measurement we make and its probability of success do not depend on the state $|\phi\rangle$. We can express the program vector in a neater form if we introduce the operator, U_{in}, defined by

$$U_{\text{in}}|00\rangle = -|10\rangle \qquad U_{\text{in}}|10\rangle = -|11\rangle$$

$$U_{\text{in}}|01\rangle = |00\rangle \qquad U_{\text{in}}|11\rangle = |01\rangle. \tag{8.27}$$

The program state can then be expressed as

$$|\Xi\rangle_{23} = \frac{1}{\sqrt{2}} U_{\text{in}}(|\phi\rangle_2|\phi_\perp\rangle_3 + |\phi_\perp\rangle_2|\phi\rangle_3). \tag{8.28}$$

Summarizing, with this device we can implement a phase flip in any basis, and the basis itself is specified by the program state.

Now let us return to our simple Controlled-NOT processor. Suppose that we want to increase the probability of a successful outcome. One possibility is to try again if we get the wrong result of our measurement on the program state. If we obtained the result $|1\rangle$ from our measurement, then the data qubit is in the state $U^{-1}(\alpha)|\psi\rangle$. We can take this qubit and run it through the processor again, but this time use the program $|\Xi(2\alpha)\rangle$. If we do so, the output state is

$$|\Psi'_{out}\rangle = \frac{1}{\sqrt{2}}(U(\alpha)|\psi\rangle|0\rangle + U^{-1}(3\alpha)|\psi\rangle|1\rangle). \tag{8.29}$$

We again measure the program state and keep the result if we get $|0\rangle$. This again happens with a probability of $1/2$. Adding this second step has increased our overall success probability to $3/4$, and the procedure can be repeated to bring the success probability as close to one as we wish. What we need to do, however, is to collect qubits in the proper program states, that is, besides a qubit in the state $|\Xi(\alpha)\rangle$, we need an additional one in the state $|\Xi(2\alpha)\rangle$.

We can also accomplish the same thing by enlarging our program space. Our data space still consists of one qubit, but the program space now contains two qubits. Let us label the three inputs: input 1 being the data input, input 2 the first program input, and input 3 the second program input. The processor now consists of two gates. The first is a controlled-NOT gate whose control qubit is qubit 1 and whose target qubit is qubit 2. The second gate is a Toffoli gate. This gate has two control qubits and one target qubit. The states of the control qubits are not changed, and if they are in the states $|0\rangle|0\rangle$, $|0\rangle|1\rangle$, or $|1\rangle|0\rangle$, neither is the state of the target qubit. However, if they are in the state $|1\rangle|1\rangle$, then σ_x is applied to the target qubit. In our processor, qubits 1 and 2 are the control qubits and qubit 3 is the target qubit. The input state is $|\psi\rangle_1|\Xi(\alpha)\rangle_2|\Xi(2\alpha)\rangle_3$, and the output state is

$$|\Psi''_{out}\rangle = \frac{1}{2}[U(\alpha)|\psi\rangle_1(|0\rangle_2|0\rangle_3 + |0\rangle_2|1\rangle_3 + |1\rangle_2|0\rangle_3) + U^{-1}(3\alpha)|\psi\rangle|1\rangle_2|1\rangle_3]. \tag{8.30}$$

At the output we measure the program qubits in the computational basis and keep the data state output if we get $|0\rangle|0\rangle$, $|0\rangle|1\rangle$, or $|1\rangle|0\rangle$. If we do, the data output is in the state $U(\alpha)|\psi\rangle$, and we have achieved our goal. This happens with a probability of $3/4$. By increasing the dimension of the program space further, we can increase our probability of success. We have, therefore, two strategies for increasing the success probability for a probabilistic processor.

8.5   A Programmable State Discriminator

In a previous chapter, we discussed unambiguous state discrimination. One is given a qubit, which is in one of two known states, $|\psi_1\rangle$ or $|\psi_2\rangle$, and one's task is to determine which of the two states the qubit is in. In the case of unambiguous discrimination, one cannot make a mistake, but the procedure is allowed to fail. We found a POVM that optimally accomplishes this task. It has three outcomes: the state is $|\psi_1\rangle$, the state is $|\psi_2\rangle$, and failure. The optimal POVM is the one that minimizes the probability of failure.

The actual state-distinguishing device, a realization of the optimal POVM, for two *known* states depends on the two states, $|\psi_1\rangle$ and $|\psi_2\rangle$, i.e., these two states are "hard wired" into the machine. What we now wish to do is to see if we can construct a machine in which the information about $|\psi_1\rangle$ and $|\psi_2\rangle$ is supplied in the form of a program. In particular, we want the program to consist of the two-qubit states that we wish to distinguish. In other words, we are given two qubits: one in the state $|\psi_1\rangle$ and another in the state $|\psi_2\rangle$. We have no knowledge of the states $|\psi_1\rangle$ and $|\psi_2\rangle$. Then we are given a third qubit that is guaranteed to be in one of these two program states, and our task is to determine, as best we can, in which one. We are allowed to fail, but not to make a mistake.

In order to solve this problem, what we need to do is to find a POVM, and our task is then reduced to the following measurement optimization problem. One has two input states:

$$|\Psi_1^{in}\rangle = |\psi_1\rangle_A |\psi_2\rangle_B |\psi_1\rangle_C,$$
$$|\Psi_2^{in}\rangle = |\psi_1\rangle_A |\psi_2\rangle_B |\psi_2\rangle_C, \tag{8.31}$$

where the subscripts A and B refer to the program registers (A contains $|\psi_1\rangle$ and B contains $|\psi_2\rangle$), and the subscript C refers to the data register. Our goal is to unambiguously distinguish between these inputs, keeping in mind that one has no knowledge of $|\psi_1\rangle$ and $|\psi_2\rangle$. In particular, one wants to find a POVM that will accomplish this.

Let the elements of our POVM be Π_1, corresponding to unambiguously detecting $|\Psi_1^{in}\rangle$, Π_2, corresponding to unambiguously detecting $|\Psi_2^{in}\rangle$, and Π_0, corresponding to failure. The probabilities of successfully identifying the two possible input states are given by

$$\langle\Psi_1^{in}|\Pi_1|\Psi_1^{in}\rangle = p_1 \qquad \langle\Psi_2^{in}|\Pi_2|\Psi_2^{in}\rangle = p_2, \tag{8.32}$$

and the condition of no errors implies that

$$\Pi_2|\Psi_1^{in}\rangle = 0 \qquad \Pi_1|\Psi_2^{in}\rangle = 0. \tag{8.33}$$

In addition, because the alternatives represented by the POVM exhaust all possibilities, we have that

$$I = \Pi_1 + \Pi_2 + \Pi_0. \tag{8.34}$$

The fact that we know nothing about $|\psi_1\rangle$ and $|\psi_2\rangle$ means that the only way we can guarantee satisfying the above conditions is to take advantage of the symmetry properties of the states, i.e., that $|\Psi_1^{in}\rangle$ is invariant under interchange of the first and third qubits and $|\Psi_2^{in}\rangle$ is invariant under interchange of the second and third qubits. That means that Π_1 should give zero when acting on states that are symmetric in qubits B and C, while Π_2 should give zero when acting on states that are symmetric in qubits A and C. Defining the antisymmetric states for the corresponding pairs of qubits

$$\left|\psi_{BC}^{(-)}\right\rangle = \frac{1}{\sqrt{2}}(|0\rangle_B|1\rangle_C - |1\rangle_B|0\rangle_C),$$

$$\left|\psi_{AC}^{(-)}\right\rangle = \frac{1}{\sqrt{2}}(|0\rangle_A|1\rangle_C - |1\rangle_A|0\rangle_C), \qquad (8.35)$$

we introduce the projectors onto the antisymmetric subspaces of the corresponding qubits as

$$P_{BC}^{as} = \left|\psi_{BC}^{(-)}\right\rangle\left\langle\psi_{BC}^{(-)}\right|,$$

$$P_{AC}^{as} = \left|\psi_{AC}^{(-)}\right\rangle\left\langle\psi_{AC}^{(-)}\right|. \qquad (8.36)$$

We can now take for Π_1 and Π_2 the operators

$$\Pi_1 = c_1 I_A \otimes P_{BC}^{as},$$

$$\Pi_2 = c_2 I_B \otimes P_{AC}^{as}, \qquad (8.37)$$

where I_A and I_B are the identity operators on the spaces of qubits A and B, respectively, and c_1 and c_2 are as yet undetermined nonnegative real numbers. Using the above expressions for Π_j, where $j = 1, 2$ in Eq. (8.32), we find that

$$p_j = \langle\Psi_j^{in}|\Pi_j|\Psi_j^{in}\rangle = c_j\frac{1}{2}(1 - |\langle\psi_1|\psi_2\rangle|^2). \qquad (8.38)$$

The average probability, P, of successfully determining which state we have, assuming that the input states occur with equal probability is given by

$$P = \frac{1}{2}(p_1 + p_2) = \frac{1}{4}(c_1 + c_2)(1 - |\langle\psi_1|\psi_2\rangle|^2), \qquad (8.39)$$

and we want to maximize this expression subject to the constraint that $\Pi_0 = I - \Pi_1 - \Pi_2$ is a positive operator.

Let S be the 4-dimensional subspace of the entire eight-dimensional Hilbert space of the three qubits, A, B, and C, that is spanned by the vectors $|0\rangle_A|\psi_{BC}^{(-)}\rangle$, $|1\rangle_A|\psi_{BC}^{(-)}\rangle$, $|0\rangle_B|\psi_{AC}^{(-)}\rangle$, and $|1\rangle_B|\psi_{AC}^{(-)}\rangle$. In the orthogonal complement of S, $S^\perp$, the operator Π_0 acts as the identity, so that in $S^\perp$, Π_0 is positive. Therefore, we need to investigate its action in S. First, let us construct an orthonormal basis for S. Applying the Gram-Schmidt process to the four vectors, given above, that span S, we obtain the orthonormal basis

$$|\Phi_1\rangle = |0\rangle_A \left|\psi_{BC}^{(-)}\right\rangle,$$

$$|\Phi_2\rangle = \frac{1}{\sqrt{3}} \left(2|0\rangle_B \left|\psi_{AC}^{(-)}\right\rangle - |0\rangle_A \left|\psi_{BC}^{(-)}\right\rangle\right),$$

$$|\Phi_3\rangle = |1\rangle_A \left|\psi_{BC}^{(-)}\right\rangle,$$

$$|\Phi_4\rangle = \frac{1}{\sqrt{3}} \left(2|1\rangle_B \left|\psi_{AC}^{(-)}\right\rangle - |1\rangle_A \left|\psi_{BC}^{(-)}\right\rangle\right). \tag{8.40}$$

In this basis, the operator Π_0, restricted to the subspace S, is given by the 4×4 matrix

$$\Pi_0 = \begin{pmatrix} 1 - c_1 - \frac{1}{4}c_2 & -\frac{\sqrt{3}}{4}c_2 & 0 & 0 \\ -\frac{\sqrt{3}}{4}c_2 & 1 - \frac{3}{4}c_2 & 0 & 0 \\ 0 & 0 & 1 - c_1 - \frac{1}{4}c_2 & -\frac{\sqrt{3}}{4}c_2 \\ 0 & 0 & -\frac{\sqrt{3}}{4}c_2 & 1 - \frac{3}{4}c_2 \end{pmatrix} \tag{8.41}$$

Because of the block diagonal nature of Π_0, the characteristic equation for its eigenvalues, λ, is given by the biquadratic equation

$$[\lambda^2 - (2 - c_1 - c_2)\lambda + 1 - (2 - c_1 - c_2) + \frac{3}{4}c_1 c_2]^2 = 0. \tag{8.42}$$

It is easy to obtain the eigenvalues explicitly, but for our purposes, the conditions that guarantee that they are nonnegative are more useful. These can be read out from the above equation, yielding

$$2 - c_1 - c_2 \geq 0,$$

$$1 - (2 - c_1 - c_2) + \frac{3}{4}c_1 c_2 \geq 0. \tag{8.43}$$

The second is the stronger of the two conditions. When it is satisfied the first one is always met, but the first one can still be used to eliminate nonphysical solutions. We can use the second condition to express c_2 in terms of c_1,

$$c_2 \leq \frac{2 - 2c_1}{2 - (3/2)c_1}. \tag{8.44}$$

For the maximum probability of success, we chose the equal sign. Inserting the resulting expression into (8.39) gives

$$P = \frac{1}{4} \left(c_1 + \frac{2 - 2c_1}{2 - (3/2)c_1}\right)(1 - |\langle\psi_1|\psi_2\rangle|^2). \tag{8.45}$$

We can easily find $c_1 = c_{1,\text{opt}}$ for which the right-hand side of this expression is maximum and using this together with Eq. (8.44) we obtain

$$c_{1,\text{opt}} = c_{2,\text{opt}} = \frac{2}{3}. \tag{8.46}$$

These values, in conjunction with Eq. (8.37), completely specify the POVM. Inserting these optimal values into (8.39) gives

$$P_{\text{POVM}} = \frac{1}{3}(1 - |\langle \psi_1 | \psi_2 \rangle|^2). \tag{8.47}$$

If we know the states $|\psi_1\rangle$ and $|\psi_2\rangle$, the probability of successfully determining the state is $1 - |\langle \psi_1 | \psi_2 \rangle|$. This is always greater than or equal to the probability in the previous equation, but this is to be expected. Knowledge of the states $|\psi_1\rangle$ and $|\psi_2\rangle$ corresponds to being given an infinite number of examples of each state, which we can then measure to determine exactly what the states are. Our programmable device only has access to one example of each state. It is, however, a very flexible device. Note that the POVM elements do not in any way depend on the states $|\psi_1\rangle$ and $|\psi_2\rangle$, which means that it will work for any two program states.

8.6 Problems

1. If we restrict the class of states that we would like to clone, we can achieve higher fidelities for the clones than is possible with a device that is designed to clone all states optimally. An example of this is phase-covariant cloning. Suppose we want to clone only states of the form

$$|\psi(\theta)\rangle = \frac{1}{\sqrt{2}}(|0\rangle + e^{i\theta}|1\rangle).$$

Consider the following cloning transformation, U, acting on two qubits

$$U|0\rangle_1|0\rangle_2 = |0\rangle_1|0\rangle_2$$
$$U|1\rangle_1|0\rangle_2 = \cos\eta|1\rangle_1|1\rangle_2 + \sin\eta|0\rangle_1|1\rangle_2.$$

The input state to this cloner is $|\psi(\theta)\rangle_1|0\rangle_2$, and the angle η controls how the information about the input state is split between outputs 1 and 2. Find the reduced density matrices of the outputs 1 and 2, the fidelities of these outputs to the input state, $|\psi(\theta)\rangle$, and show that in the case $\eta = \pi/4$ these fidelities exceed $5/6$.

2. Consider the cloner discussed in Sect. 8.2. Show that the fidelities of the output states [see Eq. (8.4)] satisfy the relation

$$\sqrt{(1-F_1)(1-F_2)} = F_1 + F_2 - \frac{3}{2},$$

where F_1 is the fidelity of the output state in output 1 to the input state and F_2 is the fidelity of the output state in output 2 to the input state.

3. Let us again consider our cloning circuit composed of four C-NOT gates. Show that it performs the following transformations:

$$|\psi\rangle_1 |0\rangle_2 |-x\rangle_3 \rightarrow (\sigma_z |\psi\rangle_2)|\Psi_-\rangle_{13}$$

$$|\psi\rangle_1 |+x\rangle_2 |1\rangle_3 \rightarrow (\sigma_x |\psi\rangle_3)|\Phi_+\rangle_{12}$$

Now show that if the input state is $|\psi\rangle_1(\alpha|\Psi_+\rangle_{23} + \beta|0\rangle_2|-x\rangle_3 + \gamma|+x\rangle_2|1\rangle_3)$, with the normalization condition

$$|\alpha + \beta|^2 + |\alpha + \gamma|^2 + |\beta - \gamma|^2 = 2,$$

the reduced density matrices of the outputs are

$$\rho_1 = \left[\left| \alpha + \frac{\beta + \gamma}{2} \right|^2 - \frac{|\beta - \gamma|^2}{4} \right] \rho_{\text{in}} + \frac{|\beta - \gamma|^2}{2} I$$

$$\rho_2 = \left[\left| \beta + \frac{(\alpha - \gamma)}{2} \right|^2 - \frac{|\alpha + \gamma|^2}{4} \right] \sigma_z \rho_{\text{in}} \sigma_z + \frac{|\alpha + \gamma|^2}{2} I$$

$$\rho_3 = \left[\left| \gamma + \frac{(\alpha - \beta)}{2} \right|^2 - \frac{|\alpha + \beta|^2}{4} \right] \sigma_x \rho_{\text{in}} \sigma_x + \frac{|\alpha + \beta|^2}{2} I,$$

where $\rho_{\text{in}} = |\psi\rangle\langle\psi|$. This implies that the cloner can not only split quantum information, but it can split it and then cause operations to be performed on the parts.

4. Suppose we want to use the probabilistic processor composed of four C-NOT gates to implement the operation $V_\phi = |\phi\rangle\langle\phi_\perp| + |\phi_\perp\rangle\langle\phi|$ on the data state. Find a program state that will cause this to happen with a probability of $1/3$, and show that if the program state is expressed in the form $U_{\text{in}}|\Xi'\rangle_{23}$, then the state $|\Xi'\rangle_{23}$ can be expressed very simply in terms of $|\phi\rangle$ and $|\phi_\perp\rangle$.

References

1. V. Bužek, M. Hillery, Quantum copying: Beyond the no-cloning theorem. Phys. Rev. A **54**, 1844 (1996)
2. V. Bužek, M. Hillery, R.F. Werner, Optimal manipulations with qubits: Universal NOT gate. Phys. Rev. A **60**, R2626 (1999)

3. For reviews of quantum cloning see V. Scarani, S. Iblisdir, N. Gisin, A. Acin, Quantum cloning. Rev. Mod. Phys. **77**, 1225 (2005); and N.J. Cerf J. Fiurašek, Optical quantum cloning—a review. Progress Opt. **49**, 455 (2006).

4. Michael A. Nielsen, Isaac L. Chuang, Programmable quantum gate arrays. Phys. Rev. Lett. **79**, 321 (1997)

5. M. Hillery, V. Bužek, M. Ziman, Probabilistic implementation of universal quantum processors. Phys. Rev. A **65**, 022301 (2002)

6. J. Preskill, Reliable quantum computers. Proc. Roy. Soc. Lond. A **454**, 385 (1998)

7. G. Vidal, L. Masanes, J.I. Cirac, Storing quantum dynamics in quantum states: stochastic programmable gate for U(1) operations. Phys. Rev. Lett. **88**, 047905 (2002)

8. J. Bergou, M. Hillery, A universal programmable quantum state discriminator that is optimal for unambiguously distinguishing between unknown states. Phys. Rev. Lett. **94**, 160501 (2005)

Chapter 9
Decoherence and Quantum Error Correction

One of the biggest problems in building a quantum computer is noise or decoherence. Qubits are coupled to other systems whether we want them to be or not, e.g., atoms couple to the electromagnetic field and spins couple to other spins via dipole–dipole interactions. These unwanted couplings can cause errors, and we need to protect quantum information against these errors.

In most of this chapter, we will study quantum error-correcting codes. These allow us to protect quantum information from the effects of decoherence. We will begin with a discussion of the general theory of quantum error-correcting codes and then discuss in detail one particular class of these codes, the Calderbank-Shor-Steane (CSS) codes. We will conclude with a very short introduction to another technique for protecting quantum information from decoherence, decoherence-free subspaces.

9.1 General Theory of Quantum Error-Correcting Codes

Classically, to protect against errors, we can just repeat the bit. We can encode one bit in three as $0 \to 000$ and $1 \to 111$. Errors can flip bits, that is, change a 0 to a 1 or vice versa. To decode the bit, we use majority voting; if there are more 0's than 1's, we call it 0, and if there are more 1's than 0's, we call it 1. This will protect against one bit-flip error.

Let's look at this in terms of probabilities. Suppose that the probability of one bit-flip error is p and that the occurrence of errors in the different bits is independent. Then the probability of no errors is $(1-p)^3$, of one error $3p(1-p)^2$, of two errors $3p^2(1-p)$, and of three errors p^3. The probability that the error correction fails is just the sum of the probabilities that two or three errors occur, or $p^2(3-2p)$. This will be smaller than the probability of an error in an unencoded bit if $p^2(3-2p) < p$, which is true if $p < 1/2$. If this condition is satisfied, then it is better to encode the bit than not.

J.A. Bergou and M. Hillery, *Introduction to the Theory of Quantum Information Processing*, Graduate Texts in Physics, DOI 10.1007/978-1-4614-7092-2_9,
© Springer Science+Business Media New York 2013

Table 9.1 Truth table for the
operations Z_1Z_2 and Z_2Z_3 on
either $|000\rangle$ or $|111\rangle$

	Z_1Z_2	Z_2Z_3
No flips	1	1
Bit 1 flipped	-1	1
Bit 2 flipped	-1	-1
Bit 3 flipped	1	-1

We would like to do something similar for qubits, but we face several problems in doing so. The task we would like to accomplish is harder, because we do not just want to protect $|0\rangle$ and $|1\rangle$, but any state of the form $a|0\rangle + b|1\rangle$. Some of the problems we face in protecting qubit states are:

1. Qubits are susceptible to more kinds of errors than are classical bits. There are phase errors that send $|0\rangle \to |0\rangle$ and $|1\rangle \to -|1\rangle$, which has the effect of changing $a|0\rangle + b|1\rangle$ to $a|0\rangle - b|1\rangle$. In addition, there are general small errors that have the effect $a|0\rangle + b|1\rangle \to (a + O(\varepsilon))|0\rangle + (b + O(\varepsilon))|1\rangle$, where $\varepsilon \ll 1$ is a parameter that characterizes the size of the error.
2. We have to be very careful about how we look at a qubit to detect the error, because by looking at a state, we mean measuring it, and measuring a state can change it.
3. We cannot just copy the qubit state, because of the no-cloning theorem.

What this means is that we have to be careful and clever.

The first quantum error-correcting code was due to Peter Shor, and we will examine it in detail. We start by analogy with the classical case and encode $|0\rangle$ by $|000\rangle$ and $|1\rangle$ by $|111\rangle$, which means that the state $a|0\rangle + b|1\rangle$ will be encoded as $a|000\rangle + b|111\rangle$. We would like to see if this encoding will help detect and correct bit-flip errors. Note that any single-qubit state is mapped into the subspace of three-qubit states spanned by $|000\rangle$ and $|111\rangle$.

If we just measure each qubit in the $\{|0\rangle, |1\rangle\}$ basis, to detect a bit-flip, we will destroy any superpositions, so something else is required. Notice that in the states $|000\rangle$ and $|111\rangle$, all of the qubits are in the same state, in particular, qubits 1 and 2 are in the same state and qubits 2 and 3 are in the same state. Denoting σ_z by Z (we will also denote σ_x by X), let us measure Z_1Z_2 and Z_2Z_3 and see what happens. Acting on either $|000\rangle$ or $|111\rangle$, we have summarized the results in Table 9.1.

So, by looking at the result, we can tell which bit flipped. In addition, any state of the form $a|000\rangle + b|111\rangle$, or this state with a single bit flipped, is an eigenstate of Z_1Z_2 and Z_2Z_3, so measuring them does not change the state. Therefore, if one bit flips, we can determine which one it is by measuring these two observables, and we will not change the state. We can then correct the error by flipping that bit back. For example, if bit 2 flipped, we would have $a|000\rangle + b|111\rangle \to a|010\rangle + b|101\rangle$; measuring Z_1Z_2 and Z_2Z_3 would give us -1 and -1, telling that it was bit 2 that flipped and not altering the state. We could then apply X_2 to the state to flip bit 2 back to its proper value.

This procedure also works if there is only some amplitude for one bit to flip. Suppose

$$|000\rangle \rightarrow (1 - \varepsilon^2)^{1/2}|000\rangle + \varepsilon|010\rangle$$
$$|111\rangle \rightarrow (1 - \varepsilon^2)^{1/2}|111\rangle + \varepsilon|101\rangle, \tag{9.1}$$

which implies that

$$a|000\rangle + b|111\rangle \rightarrow (1 - \varepsilon^2)^{1/2}(a|000\rangle + b|111\rangle) + \varepsilon(a|010\rangle + b|101\rangle). \tag{9.2}$$

Now let us see what happens when we make our measurements. Measuring $Z_1 Z_2$, we obtain 1 with probability $1 - \varepsilon^2$ and -1 with probability ε^2. If we obtain 1, the state is restored and becomes $a|000\rangle + b|111\rangle$. If we obtain -1, the state is $a|010\rangle + b|101\rangle$. Now let's measure $Z_2 Z_3$. If we obtained 1 for the first measurement, we will also obtain one for the second, since the state is now restored to what it should be. In that case, we have obtained 1 for both measurements, so we do nothing. If we obtained -1 for the first measurement, we will obtain -1 for the second, since bit 2 is definitely flipped. Having obtained -1 for both measurements, we apply X_2 to correct the error.

At this point, we can correct one bit-flip error, but now, we need to worry about phase-flip errors. Phase-flip errors behave like bit-flip errors if we look at them in a different basis. Note that a phase-flip error turns the state $|+x\rangle$ into $|-x\rangle$ and $|-x\rangle$ into $|+x\rangle$, which is the same effect a bit-flip error has in the basis $\{|0\rangle, |1\rangle\}$. If we encode $|0\rangle \rightarrow |+x, +x, +x\rangle$ and $|1\rangle \rightarrow |-x, -x, -x\rangle$, then we can detect single-bit phase-flip errors. To detect the error, we measure $X_1 X_2$ and $X_2 X_3$, which tells us in which bit the error occurred. We then correct the error by applying Z to the appropriate bit.

We now want to combine the bit- and phase-flip codes, so that we can protect against both kinds of errors. Think of starting with the phase-flip code and encoding each of the qubits in it with the bit-flip code. This gives us a nine-qubit code, which is the Shor code. In detail, the encoding is given by

$$|0\rangle \rightarrow \frac{1}{2\sqrt{2}}(|000\rangle + |111\rangle)(|000\rangle + |111\rangle)(|000\rangle + |111\rangle)$$

$$|1\rangle \rightarrow \frac{1}{2\sqrt{2}}(|000\rangle - |111\rangle)(|000\rangle - |111\rangle)(|000\rangle - |111\rangle). \tag{9.3}$$

We can find bit-flip errors by measuring products of Z operators. In particular, measuring $Z_1 Z_2$ and $Z_2 Z_3$ will detect bit-flip errors in the first three-qubit cluster, measuring $Z_4 Z_5$ and $Z_5 Z_6$ will detect bit-flip errors in the second three-qubit cluster, and $Z_7 Z_8$ and $Z_8 Z_9$ will detect bit-flip errors in the third three-qubit cluster. Once the bit-flip has been detected, we can apply an X operator to the appropriate qubit to flip it back.

A phase-flip error in any qubit will cause the sign in one of the clusters to flip. We can find which cluster by measuring $\prod_{j=1}^{6} X_j$ and $\prod_{j=4}^{9} X_j$. If both give 1, then there is no error, if the first gives 1 and the second -1, the error is in the first cluster, if both give -1, the error is in the second cluster, and if the first gives -1 and the second 1, the error is in the third cluster. Once we have determined in which cluster the error occurred, we can apply a Z operator to any of the qubits in that cluster to correct the error. Note that this code will correct an error in one qubit, but not more.

So far we have only considered bit-flip and phase-flip errors. It doesn't seem as though this would be sufficient, but it is. To see why, we must take a more general look at quantum error correction. We start by considering a single qubit interacting with its environment. Let the qubit Hilbert space be $\mathcal{H}_A$ and the environment Hilbert space be $\mathcal{H}_E$. We shall call the initial state of the environment $|0\rangle_E$ and the operator that describes the evolution of the qubit and the environment U_{AE}. We have that

$$U_{AE}(|0\rangle_A \otimes |0\rangle_E) = |0\rangle_A \otimes |e_{00}\rangle_E + |1\rangle_A \otimes |e_{01}\rangle_E$$
$$U_{AE}(|1\rangle_A \otimes |0\rangle_E) = |0\rangle_A \otimes |e_{10}\rangle_E + |1\rangle_A \otimes |e_{11}\rangle_E. \tag{9.4}$$

The states $|e_{jk}\rangle_E$ are not necessarily orthogonal or normalized, but they must obey the constraints imposed by the unitarity of U_{AE}. For example, we must have that $\|e_{00}\|^2 + \|e_{01}\|^2 = 1$ and $\|e_{10}\|^2 + \|e_{11}\|^2 = 1$. We now want to see the effect of U_{AE} acting on a general qubit state, i.e., on $|\psi\rangle_A \otimes |0\rangle_E$, where $|\psi\rangle_A = a|0\rangle_A + b|1\rangle_A$. After some work we find that

$$U_{AE}(|\psi\rangle_A \otimes |0\rangle_E) = a(|0\rangle_A \otimes |e_{00}\rangle + |1\rangle_A \otimes |e_{01}\rangle_E)$$
$$+ b(|0\rangle_A \otimes |e_{10}\rangle_E + |1\rangle_A \otimes |e_{11}\rangle_E)$$
$$= I|\psi\rangle_A \otimes |e_I\rangle_E + X|\psi\rangle_A \otimes |e_X\rangle_E$$
$$+ Y|\psi\rangle_A \otimes |e_Y\rangle_E + Z|\psi\rangle_A \otimes |e_Z\rangle_Z, \tag{9.5}$$

where I is the identity operator, $Y = iXZ$, and

$$|e_I\rangle_E = \frac{1}{2}(|e_{00}\rangle + |e_{11}\rangle) \qquad |e_X\rangle_E = \frac{1}{2}(|e_{01}\rangle + |e_{10}\rangle)$$
$$|e_Y\rangle_E = \frac{i}{2}(|e_{10}\rangle - |e_{01}\rangle) \qquad |e_Z\rangle_E = \frac{1}{2}(|e_{00}\rangle - |e_{11}\rangle). \tag{9.6}$$

Therefore, we can expand the action of U_{AE} on the qubit in terms of the Pauli matrices. This is a consequence of the fact that these matrices plus the identity form basis for 2×2 matrices. Note that the vectors $|e_I\rangle_E$, $|e_X\rangle_E$, $|e_Y\rangle_E$, and $|e_Z\rangle_E$ are not necessarily normalized or orthogonal. For n qubits we can expand the unitary evolution operator that mixes the qubits and the environment in terms of $\{I, X, Y, Z\}^{\otimes n}$. Let us call the members of this set E_a so that

$$U_{AE}(|\psi\rangle_A \otimes |0\rangle_E) = \sum_a E_a |\psi\rangle_A \otimes |e_a\rangle_E. \tag{9.7}$$

Note that E_a is unitary and that $\mathcal{H}_A$ is now the n-qubit Hilbert space.

When designing a code, we choose a subset $\mathcal{E} \subseteq \{I, X, Y, Z\}^{\otimes n}$; these are the errors we want to be able to correct. Typically, $\mathcal{E}$ is chosen to be all E_a of weight t or less. The weight of E_a is the number of operators it contains that are not the identity. Next, we choose a code subspace, $\mathcal{H}_c \subseteq \mathcal{H}_A$, which will contain the code words, and suppose $\{|\bar{j}\rangle_A\}$ is an orthonormal basis of that space. Suppose we had for $E_a, E_b \in \mathcal{E}$

$$_A\langle \bar{j}|E_b^\dagger E_a|\bar{k}\rangle_A = \delta_{ab}\delta_{\bar{j}\bar{k}}. \tag{9.8}$$

This implies that each error in $\mathcal{E}$ maps the code space into a different subspace and that all of these subspaces are orthogonal, i.e., $E_a\mathcal{H}_A$ is orthogonal to $E_b\mathcal{H}_A$ for $a \neq b$, and hence these subspaces are distinguishable. Within one of these subspaces, errors map code words (the basis elements $|\bar{j}\rangle_A$) onto orthogonal states, that is, $E_a|\bar{j}\rangle_A$ is orthogonal to $E_a|\bar{k}\rangle_A$ for $\bar{j} \neq \bar{k}$.

This means that we can find which error occurred (into which orthogonal subspace it mapped the code word), and we can correct it. If we found that E_a occurred, we just apply $E_a^\dagger$. In fact, we can correct any error that is a combination of the elements of $\mathcal{E}$. If

$$|\psi\rangle_A \otimes |0\rangle_E \rightarrow \sum_a E_a|\psi\rangle_A \otimes |e_a\rangle_E, \tag{9.9}$$

then we can measure the observable $\sum_a \lambda_a P_a$, where the λ_a are distinct, and P_a projects onto $E_a\mathcal{H}_A$. If we obtain $\lambda_{a'}$, then the state becomes $E_{a'}|\psi_A \otimes |e_{a'}\rangle_E$, and we can apply $E_{a'}^\dagger$ to correct the error. Therefore, by being able to correct a finite number of errors, in particular the elements of $\mathcal{E}$, we are able to correct an infinite number of them, i.e., any combination of the errors in $\mathcal{E}$.

It turns out that the condition in Eq. (9.8) is too strong. The Shor code does not obey it, and it still works. In that code, different phase-flip errors in the same cluster lead to identical states. A code satisfying Eq. (9.8) is called a nondegenerate code. Codes that do not satisfy it are called degenerate.

Before discussing the general condition for a quantum code to correct a set of errors, let us show that both errors and the recovery process can be represented as superoperators. Let $\{|\mu\rangle_E\}$ be an orthonormal basis for $\mathcal{H}_E$. We can expand the states $|e_a\rangle_E$ appearing in Eq. (9.9) in this basis, and this allows us to express Eq. (9.9) as

$$U_{AE}|\psi\rangle_A \otimes |0\rangle_E = \sum_\mu M_\mu|\psi\rangle_A \otimes |\mu\rangle_E, \tag{9.10}$$

where

$$M_\mu = \sum_a {}_E\langle \mu|e_a\rangle_E E_a. \tag{9.11}$$

The unitarity of U_{AE} implies that $\sum_\mu M_\mu^\dagger M_\mu = I$. Tracing out the environment, we see that the error takes the density matrix in the code subspace, ρ_A, to

$$T_E(\rho_A) = \sum_\mu M_\mu \rho_A M_\mu^\dagger. \tag{9.12}$$

We see, then, that errors can be represented as superoperators.

Now let us look at the recovery process. Let ρ_A' be the state of n qubits after the error. We measure ρ_A', the measurement being described by a POVM with operators $\tilde{R}_v$, and if we get result v we apply the operator U_v to correct the error. Therefore, with probability $p_v = \mathrm{Tr}(\tilde{R}_v^\dagger \tilde{R}_v \rho_A')$, we obtain the state

$$\rho_{Av} = \frac{1}{p_v} U_v \tilde{R}_v \rho_A' \tilde{R}_v^\dagger U_v^\dagger. \tag{9.13}$$

Defining $R_v = U_v \tilde{R}_v$, we have that the entire density matrix after the correction procedure has been applied is

$$R(\rho_A') = \sum_v p_v \rho_{Av} = \sum_v R_v \rho_A' R_v^\dagger. \tag{9.14}$$

Note that

$$\sum_v R_v^\dagger R_v = \sum_v \tilde{R}_v^\dagger \tilde{R}_v = I, \tag{9.15}$$

because $\{\tilde{R}_v\}$ is a POVM, and therefore R is a superoperator.

We are now in a position to show that the condition for a quantum code to be able to correct an error described by the superoperator T_E, which has Kraus operators M_μ, is

$$_A\langle \bar{j} | M_{\mu'}^\dagger M_\mu | \bar{k} \rangle_A = C_{\mu'\mu} \delta_{\bar{j}\bar{k}}, \tag{9.16}$$

for all M_μ and $M_{\mu'}$, where $C_{\mu'\mu}$ is an arbitrary hermitian matrix. In order to analyze this claim, we will work on an extended space $\mathcal{H}_A \otimes \mathcal{H}_E \otimes \mathcal{H}_B$, which will allow us to use state vectors instead of density matrices. On this space T_E can be represented as $U_{AE} \otimes I_B$, that is, a unitary operator that acts on $\mathcal{H}_A \otimes \mathcal{H}_E$ and the identity on $\mathcal{H}_B$, and R can be represented as $U_{AB} \otimes I_E$, that is, a unitary operator that acts on $\mathcal{H}_A \otimes \mathcal{H}_B$ and the identity on $\mathcal{H}_E$. In detail we have

$$T_E : |\bar{j}\rangle_A \otimes |0\rangle_E \otimes |v\rangle_B \to \sum_\mu M_\mu |\bar{j}\rangle_A \otimes |\mu\rangle_E \otimes |v\rangle_B$$

$$R : |\bar{j}\rangle_A \otimes |\mu\rangle_E \otimes |0\rangle_B \to \sum_v R_v |\bar{j}\rangle_A \otimes |\mu\rangle_E \otimes |v\rangle_B. \tag{9.17}$$

If the recovery operation is to correct the error on the code subspace, we must have

$$R \circ T_E : |\bar{j}\rangle_A \otimes |0\rangle_E \otimes |0\rangle_B \to \sum_{\mu,v} R_v M_\mu |\bar{j}\rangle_A \otimes |\mu\rangle_E \otimes |v\rangle_B = |\bar{j}\rangle_A \otimes |\Psi\rangle_{EB}, \tag{9.18}$$

where $|\Psi\rangle_{EB}$ is independent of $\bar{j}$. Taking the inner product of both sides with $_E\langle \mu' |_B\langle v'|$, we have that

$$R_{v'} M_{\mu'} |\bar{j}\rangle_A = \lambda_{\mu'v'} |\bar{j}\rangle_A, \tag{9.19}$$

where $\lambda_{\mu'\nu'} = {}_E\langle\mu'|_B\langle\nu'|\Psi\rangle_{EB}$ is independent of $\bar{j}$. This implies that for any $|\psi\rangle_A$ in the code space, $R_\nu M_\mu|\psi\rangle_A = \lambda_{\mu\nu}|\psi\rangle_A$, so that for $|\phi\rangle_A$ in the code space

$$ {}_A\langle\phi|R_\nu M_\mu|\psi\rangle_A = \lambda_{\mu\nu}\,{}_A\langle\phi|\psi\rangle_A = {}_A\langle(R_\nu M_\mu)^\dagger\phi|\psi\rangle_A. \tag{9.20} $$

This further implies that $(R_\nu M_\mu)^\dagger|\phi\rangle_A = \lambda_{\mu\nu}^*|\phi\rangle_A$ for $|\phi\rangle_A$ in the code space. We now have

$$ M_\sigma^\dagger M_\mu|\bar{j}\rangle_A = M_\sigma^\dagger(\sum_\nu R_\nu^\dagger R_\nu)M_\mu|\bar{j}\rangle_A = (\sum_\nu \lambda_{\sigma\nu}^*\lambda_{\mu\nu})|\bar{j}\rangle_A. \tag{9.21} $$

so that, setting $C_{\sigma\mu} = \sum_\nu \lambda_{\sigma\nu}^*\lambda_{\mu\nu}$,

$$ {}_A\langle\bar{k}|M_\sigma^\dagger M_\mu|\bar{j}\rangle_A = C_{\sigma\mu}\delta_{\bar{k}\bar{j}}. \tag{9.22} $$

Therefore, we have shown that if the recovery operation is able to correct the error, the condition in Eq. (9.16) must be satisfied.

Now let us show the reverse, if the condition in Eq. (9.16) is satisfied, then we can recover from the error produced by T_E. First, let us define a new Kraus representation for T_E by

$$ \tilde{M}_\mu = \sum_{\mu'} u_{\mu\mu'} M_{\mu'}, \tag{9.23} $$

where $u_{\mu\mu'}$ is a unitary matrix. This gives us

$$ {}_A\langle\bar{k}|\tilde{M}_\sigma^\dagger \tilde{M}_\mu|\bar{j}\rangle_A = \delta_{\bar{k}\bar{j}}\sum_{\sigma'\mu'} u_{\sigma\sigma'}^* C_{\sigma'\mu'} u_{\mu\mu'} = \delta_{\bar{k}\bar{j}}\sum_{\sigma'\mu'} u_{\sigma\sigma'}^* C_{\sigma'\mu'} (u^*)_{\mu'\mu}^\dagger. \tag{9.24} $$

We can now choose u^* to diagonalize C, so that the above equation becomes

$$ {}_A\langle\bar{k}|\tilde{M}_\sigma^\dagger \tilde{M}_\mu|\bar{j}\rangle_A = \delta_{\bar{k}\bar{j}}\tilde{C}_\mu\delta_{\sigma\mu}. \tag{9.25} $$

Note that because $\sum_\mu \tilde{M}_\mu^\dagger \tilde{M}_\mu = I$, we have that $\sum_\mu \tilde{C}_\mu = 1$. For each $\tilde{C}_\nu \neq 0$ define

$$ R_\nu = \frac{1}{\sqrt{\tilde{C}_\nu}}\sum_{\bar{k}} |\bar{k}\rangle_A\langle\bar{k}|\tilde{M}_\nu^\dagger. \tag{9.26} $$

First we note that

$$ R_\nu\tilde{M}_\mu|\bar{j}\rangle_A = \frac{1}{\sqrt{\tilde{C}_\nu}}\sum_{\bar{k}}|\bar{k}\rangle_A\langle\bar{k}|\tilde{M}_\nu^\dagger \tilde{M}_\mu|\bar{j}\rangle_A = \sqrt{\tilde{C}_\nu}\,\delta_{\mu\nu}|\bar{j}\rangle_A. \tag{9.27} $$

Going back to our representation of the superoperators on $\mathcal{H}_A \otimes \mathcal{H}_E \otimes \mathcal{H}_B$, we have that

$$ \sum_{\mu\nu} R_\nu\tilde{M}_\mu|\bar{j}\rangle_A \otimes |\mu\rangle_E \otimes |\nu\rangle_B = |\bar{j}\rangle_A \otimes \sum_\mu \sqrt{\tilde{C}_\nu}|\mu\rangle_E \otimes |\mu\rangle_B = |\bar{j}\rangle_A|\Psi\rangle_{EB}, \tag{9.28} $$

so that it does recover the original state in the code space. Finally, we need to verify that $\sum_v R_v^\dagger R_v = I$. We begin by noting that

$$\sum_v R_v^\dagger R_v = \sum_v \sum_{\bar{j}} \frac{1}{\tilde{C}_v} \tilde{M}_v |\bar{j}\rangle_A \langle \bar{j}| \tilde{M}_v^\dagger. \tag{9.29}$$

Now let us apply this operator to any vector of the form $\tilde{M}_\sigma |\bar{k}\rangle_A$,

$$\sum_v R_v^\dagger R_v \tilde{M}_\sigma |\bar{k}\rangle_A = \sum_v \sum_{\bar{j}} \frac{1}{\tilde{C}_v} \tilde{M}_v |\bar{j}\rangle_A \langle \bar{j}| \tilde{M}_v^\dagger \tilde{M}_\sigma |\bar{k}\rangle_A$$

$$= \sum_v \sum_{\bar{j}} \frac{1}{\tilde{C}_v} \tilde{M}_v |\bar{j}\rangle_A \tilde{C}_v \delta_{v\sigma} \delta_{\bar{j}\bar{k}} = \tilde{M}_\sigma |\bar{k}\rangle_A. \tag{9.30}$$

Defining $\mathcal{H}_{\tilde{M}} = \text{span}\{\tilde{M}_\sigma |\psi\rangle_A\}$ for all $\tilde{M}_\sigma$ and $|\psi\rangle_A \in \mathcal{H}_c$, we see that $\sum_v R_v^\dagger R_v$ is just the projection onto $\mathcal{H}_{\tilde{M}}$. To complete the recovery operation, we just add to it $P_{\tilde{M}}^\perp$, the projection onto the orthogonal complement of $\mathcal{H}_{\tilde{M}}$. Adding this operator does not affect our recovery operation, because this operation takes place in $\mathcal{H}_{\tilde{M}}$, and $P_{\tilde{M}}^\perp$ maps any state in this space to zero.

Summarizing, what we have shown is that we can recover from an error T_E with Kraus operators M_μ, if and only if Eq. (9.16) is satisfied. Now this does not appear to be too impressive; we can recover from one error. However, the situation is better than it seems. The same recovery procedure will work for any error whose Kraus operators are linear combinations of the M_μ. In order to see this consider an error T_F with Kraus operators

$$F_\sigma = \sum_\mu m'_{\sigma\mu} M_\mu = \sum_\mu m_{\sigma\mu} \tilde{M}_\mu. \tag{9.31}$$

Applying our recovery operator to a code word affected by F_σ gives us

$$R_v F_\sigma |\bar{j}\rangle_A = \frac{1}{\sqrt{\tilde{C}_v}} \sum_{\bar{k}} \sum_\mu m_{\sigma\mu} |\bar{k}\rangle_A \langle \bar{k}| \tilde{M}_v^\dagger \tilde{M}_\mu |\bar{j}\rangle_A = \sqrt{\tilde{C}_v} m_{\sigma v} |\bar{j}\rangle_A. \tag{9.32}$$

Going back to our description of the error and recovery operations on the extended space $\mathcal{H}_A \otimes \mathcal{H}_E \otimes \mathcal{H}_B$, we have

$$\sum_{v\sigma} R_v F_\sigma |\bar{j}\rangle_A \otimes |\sigma\rangle_E \otimes |v\rangle_B = |\bar{j}\rangle_A \otimes \sum_{v\sigma} \sqrt{\tilde{C}_v} m_{\sigma v} |\sigma\rangle_E \otimes |v\rangle_B = |\bar{j}\rangle_A |\Psi\rangle_{EB}, \tag{9.33}$$

so the error is corrected.

Now that we know what is necessary to correct errors, let us go back and consider the basic errors $E_a \in \mathcal{E}$ from which we built up all of the others. Define $T_{\mathcal{E}}$ to have Kraus operators $\sqrt{p_a} E_a$, where $0 \le p_a \le$ and $\sum_a p_a = 1$. Then, if and only if our code space satisfies

$$_A\langle \bar{j}| E_b^\dagger E_a |\bar{k}\rangle_A = C_{ba} \delta_{\bar{j}\bar{k}}, \tag{9.34}$$

can we recover from $T_{\mathcal{E}}$. But if we can recover from $T_{\mathcal{E}}$, then we can recover from any error whose Kraus operators are linear combinations of the $E_a \in \mathcal{E}$. For example, if $\mathcal{E}$ consists of bit flips, phase flips, or both on t qubits or fewer, then we can recover from all errors on t qubits or fewer if our code satisfies the above condition.

9.2 An Example: CSS Codes

Now we want to look at a particular class of quantum codes, the CSS codes. Before we do, however, it is necessary to learn something about classical linear codes. A linear code that encodes k bits of information into n bits is called an $[n,k]$ code. It can be described by an $n \times k$ matrix (n rows and k columns) whose elements are 0 or 1. This matrix, G, is known as the generator matrix for the code. A k-digit binary number is encoded into an n-digit code word by writing it as a column vector of length k and then multiplying this vector by the generator matrix to give a column vector of length n, which is the code word. All of the operations here are modulo 2, so the elements of the vectors and matrix are members of the field F_2 that contains the elements 0 and 1, and whose operations, addition and multiplication, are done modulo 2. As an example, consider the $[6,2]$ code with generator matrix

$$G = \begin{pmatrix} 1 & 0 \\ 1 & 0 \\ 1 & 0 \\ 0 & 1 \\ 0 & 1 \\ 0 & 1 \end{pmatrix}. \tag{9.35}$$

This encodes two bits into six as follows:

$$\begin{pmatrix} 0 \\ 0 \end{pmatrix} \rightarrow \begin{pmatrix} 0 \\ 0 \\ 0 \\ 0 \\ 0 \\ 0 \end{pmatrix}, \quad \begin{pmatrix} 1 \\ 0 \end{pmatrix} \rightarrow \begin{pmatrix} 1 \\ 1 \\ 1 \\ 0 \\ 0 \\ 0 \end{pmatrix}, \tag{9.36}$$

etc. The space of code words, C, is spanned by the columns of G. These should be linearly independent over F_2 so that the encoding is unique. Every vector in C will be a code word.

Another way to specify the code subspace is by means of constraints. Our code subspace has dimension k and lies in an n-dimensional space, so we can specify it by imposing $n-k$ constraints. This can be done by means of an $n-k$ by n matrix H. The code subspace is the set of n-component vectors that is mapped to 0 by H.

If the $n - k$ constraints are independent, then the rows of H will be independent. H is called the parity-check matrix, and, as we shall see, it is useful in correcting errors.

Clearly G and H are related. Since the columns of G are in the code subspace, we have that $HG = 0$. Let's find H for the $[6,2]$ code. It will be a 4×6 matrix, and its rows need to be orthogonal to the columns of G. That means we need four linearly independent six-component vectors that are orthogonal to the two columns of G. We start by noticing that $(1\ 1\ 0)^T$ and $(1\ 0\ 1)^T$ are linearly independent and are orthogonal to $(1\ 1\ 1)^T$, where T denotes transpose. We can, therefore, choose

$$H = \begin{pmatrix} 1 & 1 & 0 & 0 & 0 & 0 \\ 1 & 0 & 1 & 0 & 0 & 0 \\ 0 & 0 & 0 & 1 & 1 & 0 \\ 0 & 0 & 0 & 1 & 0 & 1 \end{pmatrix}. \tag{9.37}$$

The parity-check matrix is useful for detecting and correcting errors, because an error will usually send the code word out of the code subspace, and we can detect this by acting on the corrupted code word with H. To see how this works, we first define the weight of an n-component vector consisting of 0's and 1's to be the number of 1's. We can represent a corrupted code word x by $x + e$, where e is an n-component vector representing the error. Each 1 in e causes a bit-flip error in x, so that the number of bit-flip errors is equal to the weight of e. Note that because $Hx = 0$, we have that $H(x + e) = He$, and we call He the syndrome of error e. Define the distance of a code to be the minimum weight of any nonzero code word, i.e., of any nonzero $x \in C$. The Hamming distance, which we shall just call the distance, between two code words x and y is just the number of places in which they differ, which is the same as the weight of $x + y$. We shall denote this distance by $d(x,y)$. As a result of these definitions, we see that for $x \neq y$, $d(x,y)$ will be greater than or equal to the weight of the code, because since $x + y \in C$, its weight must be greater than or equal to the weight of the code. Therefore, if a code C has a distance of $2t + 1$, then errors of weight t will not change one code word into another. Each error will produce a unique syndrome so that we can correct it. To see this, note that if $e_1 \neq e_2$ but $He_1 = He_2$, then $H(e_1 + e_2) = 0$ so that we would have $e_1 + e_2 \in C$. This, however, is not possible, because the weight of $e_1 + e_2$ is less than or equal to $2t$, but the weight of the code is $2t + 1$. Therefore, $He_1 \neq He_2$, and the error syndromes are unique. Once we know which error has occurred, say e, we can correct it by adding e to the corrupted code word, because $(x + e) + e = x$.

For each code C, there is a dual code $C^\perp$. This comes from the observation that $HG = 0$ implies that $G^T H^T = 0$, so that we can interpret H^T as a generator matrix for an $[n, n - k]$ code and G^T as its parity-check matrix. This equation implies that each code word in $C^\perp$ is orthogonal to all of the columns of G, so that each code word in $C^\perp$ is orthogonal to all of the code words in C. Because vectors in F_2^n can be orthogonal to themselves, C and $C^\perp$ can intersect. A code is called weakly self-dual if $C \subseteq C^\perp$ and self-dual if $C = C^\perp$. For an $[n,k]$ code to be self-dual, we must have $n = 2k$.

In concluding our brief introduction to classical linear codes, we want to prove an identity relating C and $C^\perp$, which will come in useful shortly. The identity is

$$\sum_{x \in C} (-1)^{x \cdot y} = \begin{cases} 2^k & y \text{ in } C^\perp \\ 0 & y \text{ not in } C^\perp \end{cases} . \tag{9.38}$$

The first part is easy. If $y \in C^\perp$, and $x \in C$, then $x \cdot y = 0$. Now, using the fact that C has 2^k code words, we get the first result. The second part follows from the identity, where $w \in \{0,1\}^k$,

$$\sum_{v \in \{0,1\}^k} (-1)^{v \cdot w} = 0, \tag{9.39}$$

for $w \neq 0$. We can express $x \in C$ as $x = Gv$ for some $v \in \{0,1\}^k$, so we have that

$$\sum_{x \in C} (-1)^{x \cdot y} = \sum_{v \in \{0,1\}^k} (-1)^{(Gv) \cdot y} = \sum_{v \in \{0,1\}^k} (-1)^{v \cdot (G^T y)} = 0, \tag{9.40}$$

if $G^T y \neq 0$. But, $G^T y \neq 0$ implies that y is not in $C^\perp$.

Now we can use these classical codes to define a quantum code. Let C_1 be an $[n,k_1]$ classical code and C_2 be an $[n,k_2]$ classical code, where $k_1 > k_2$ and $C_2 \subset C_1$. We further suppose that C_1 has a distance d_1 and $C_2^\perp$ has a distance $d_2^\perp$. We define two elements of C_1, x and y, to be equivalent if and only if $x + y \in C_2$. This breaks C_1 up into $|C_1|/|C_2| = 2^{k_1 - k_2}$ equivalence classes, or cosets. We define an n-qubit quantum state for each coset as

$$|x + C_2\rangle = \frac{1}{\sqrt{|C_2|}} \sum_{y \in C_2} |x + y\rangle. \tag{9.41}$$

The fact that the cosets are disjoint means that these states are orthogonal for x and x' in different cosets. These states span a $2^{k_1 - k_2}$-dimensional subspace of the n-qubit space, so this is an $[n, k_1 - k_2]$ quantum code; it encodes $k_1 - k_2$ qubits in n qubits.

Let us see what happens when we apply $H^{\otimes n}$, a Hadamard gate to each qubit, to this state. Remember that

$$H^{\otimes n}|x\rangle = \frac{1}{2^{n/2}} \sum_{y=0}^{2^n - 1} (-1)^{x \cdot y} |y\rangle \tag{9.42}$$

so

$$H^{\otimes n}|x + C_2\rangle = \frac{1}{\sqrt{|C_2|}} \sum_{y \in C_2} \frac{1}{2^{n/2}} \sum_{u=0}^{2^n - 1} (-1)^{(x+y) \cdot u}$$

$$= \frac{1}{2^{(n+k_2)/2}} \sum_{u=0}^{2^n - 1} (-1)^{x \cdot u} \sum_{y \in C_2} (-1)^{y \cdot u} |u\rangle$$

$$= \frac{1}{2^{(n-k_2)/2}} \sum_{u \in C_2^\perp} (-1)^{x \cdot u} |u\rangle. \tag{9.43}$$

What we get is a superposition, with phases, of code words in $C_2^\perp$. As we shall see, it will be possible to correct bit-flip errors in the original code and phase-flip errors in $C_2^\perp$.

Now suppose $d_1 > 2t_f + 1$ and $d_2^\perp > 2t_p + 1$. Our code will then be able to correct t_f bit-flip errors and t_p phase-flip errors. Let e_1 be a vector with weight less than t_f, and e_2 be a vector with weight less than t_p. The ones in e_1 correspond to bit flips and the ones in e_2 correspond to phase flips. The errors have the effect

$$|x + C_2\rangle \rightarrow \frac{1}{\sqrt{|C_2|}} \sum_{y \in C_2} (-1)^{(x+y) \cdot e_2} |x + y + e_1\rangle. \tag{9.44}$$

To correct the bit-flip errors we append an n-qubit ancilla and apply a unitary operator that takes

$$U_f |v\rangle |0\rangle = |v\rangle |H_1 v\rangle, \tag{9.45}$$

where H_1 is the parity-check matrix for C_1. We then have that

$$U_f \left(\frac{1}{\sqrt{|C_2|}} \sum_{y \in C_2} (-1)^{(x+y) \cdot e_2} |x + y + e_1\rangle \right) |0\rangle$$

$$= \left(\frac{1}{\sqrt{|C_2|}} \sum_{y \in C_2} (-1)^{(x+y) \cdot e_2} |x + y + e_1\rangle \right) |H_1 e_1\rangle. \tag{9.46}$$

Now measure the ancilla in the computational basis. The result tells us which qubits have been flipped, as C_1 can correct up to t_f bit-flip errors. Apply X to these bits to flip them back, and throw away the ancilla. Our state is now

$$\frac{1}{\sqrt{|C_2|}} \sum_{y \in C_2} (-1)^{(x+y) \cdot e_2} |x + y\rangle. \tag{9.47}$$

Now apply $H^{\otimes n}$ to this state

$$H^{\otimes n} \frac{1}{\sqrt{|C_2|}} \sum_{y \in C_2} (-1)^{(x+y) \cdot e_2} |x + y\rangle$$

$$= \frac{1}{\sqrt{|C_2|}} \sum_{y \in C_2} \frac{1}{2^{n/2}} \sum_{u=0}^{2^n - 1} (-1)^{(x+y) \cdot (e_2 + u)} |u\rangle$$

$$= \frac{1}{2^{(n-k_2)/2}} \sum_{u + e_2 \in C_2^\perp} (-1)^{x \cdot (u + e_2)} |u\rangle$$

$$= \frac{1}{2^{(n-k_2)/2}} \sum_{u' \in C_2^{\perp}} (-1)^{x \cdot u'} |u' + e_2\rangle, \tag{9.48}$$

where $u' = u + e_2$. Now append an n-qubit ancilla and apply a unitary operator that takes

$$U_p |v\rangle |0\rangle = |v\rangle |G_2^T v\rangle, \tag{9.49}$$

where G_2 is the generator for C_2 so that G_2^T is the parity-check matrix for $C_2^{\perp}$. Doing so gives us

$$U_p \left(\frac{1}{2^{(n-k_2)/2}} \sum_{u' \in C_2^{\perp}} (-1)^{x \cdot u'} |u' + e_2\rangle \right)$$

$$= \left(\frac{1}{2^{(n-k_2)/2}} \sum_{u' \in C_2^{\perp}} (-1)^{x \cdot u'} |u' + e_2\rangle \right) |G_2^T e_2\rangle. \tag{9.50}$$

We again measure the ancilla in the computational basis, and this tells us which bits have flipped. Apply X to these bits to flip them back, and discard the ancilla. We now have the state

$$\frac{1}{2^{(n-k_2)/2}} \sum_{u' \in C_2^{\perp}} (-1)^{x \cdot u'} |u'\rangle. \tag{9.51}$$

Now apply $H^{\otimes n}$, and from Eq. (9.43) and the fact that $H^2 = I$, we have

$$H^{\otimes n} \frac{1}{2^{(n-k_2)/2}} \sum_{u' \in C_2^{\perp}} (-1)^{x \cdot u'} |u'\rangle = \frac{1}{\sqrt{|C_2|}} \sum_{y \in C_2} |x + y\rangle, \tag{9.52}$$

and all of the errors have been corrected. Note that if we assume that $t_f = t_p = t$, what we have shown is that we can correct t bit flips, t phase flips, and t products of bit flips and phase flips. This implies that we can correct all errors on t qubits or fewer.

An example of a CSS code is the 7-qubit Steane code, which can correct errors in one qubit. It is based on the classical $[7,4]$ Hamming code. A Hamming code is derived by choosing an integer $r \geq 2$ and then taking for the parity-check matrix, H, the matrix whose columns are the $2^r - 1$ bit strings of length r, excluding the string with all zeroes. This gives an r by $2^r - 1$ parity-check matrix, which implies the generator is a $2^r - 1$ by $2^r - r - 1$ matrix, so we have a $[2^r - 1, 2^r - r - 1]$ code. If $r = 3$, this gives a $[7,4]$ code. The parity-check matrix for this code is

$$H = \begin{pmatrix} 1 & 0 & 1 & 0 & 1 & 0 & 1 \\ 0 & 1 & 1 & 0 & 0 & 1 & 1 \\ 0 & 0 & 0 & 1 & 1 & 1 & 1 \end{pmatrix}. \tag{9.53}$$

This code has distance 3. To see this first note that the string $x_3 = (1110000)^T$, which has weight 3, satisfies $Hx_3 = 0$, so it is in the code. If x_1 were a code word of weight 1, then $Hx_1 = 0$ would imply that one of the columns of H would have to be zero, which is not the case. Therefore, there are no weight-one code words. If x_2 is a code word of weight 2, we could express it as $x_2 = x_1 + x_1'$, where x_1 and x_1' are both of weight one and $x_1 \neq x_1'$. Then $H(x_1 + x_1') = Hx_1 + Hx_1' = 0$ implies that two columns of H must be identical, which is also not the case. Therefore, there are no weight-two code words, and the distance of the code is 3. The generator matrix for this code is

$$G = \begin{pmatrix} 1 & 0 & 0 & 1 \\ 0 & 1 & 0 & 1 \\ 1 & 1 & 0 & 1 \\ 0 & 0 & 1 & 0 \\ 1 & 0 & 1 & 0 \\ 0 & 1 & 1 & 0 \\ 1 & 1 & 1 & 0 \end{pmatrix}. \tag{9.54}$$

Note that the rows of H are in the code, and they are just the first three columns of G. These vectors are orthogonal to themselves.

The matrix H^T is the generator of the dual code, which is a $[7,3]$ code. In this case $C^\perp \subset C$, and $C^\perp$ consists of all of the code words in C with an even weight. This code also has a distance of 3.

To construct the CSS code, we take $C_1 = C$ and $C_2 = C^\perp$, so that $C_2^\perp = C$ has weight 3. This implies that the Steane code can correct one-qubit errors. It is a $[7,1]$ quantum code, $(k_1 - k_2 = 4 - 3 = 1)$. There are only two cosets in this case, each with eight members. Letting y_j, for $j = 1,2,3,4$ be the columns of G, we have that the members of the coset containing the identity are given by $c_1y_1 + c_2y_2 + c_3y_3$, where $c_j \in \{0,1\}$, and the members of the other coset are given by $c_1y_1 + c_2y_2 + c_3y_3 + y_4$. Note that the members of the first coset have even weight, while the members of the second have odd weight.

9.3 Decoherence-Free Subspaces

So far we have focussed on error-correcting codes as a way of defeating the effects of decoherence. There are a number of other approaches, and we will give here a very brief introduction to one of them. This method takes advantage of the fact that if qubits are subject to the same errors, there are subspaces that remain free from the effects of decoherence.

Let us start with a single qubit and suppose it is subject to random-phase errors. In particular, we have that

$$|0\rangle \to |0\rangle \quad |1\rangle \to e^{i\phi}|1\rangle, \tag{9.55}$$

where ϕ is distributed according to probability distribution $p(\phi)$. Let us suppose that the qubit is initially in the state $|\psi\rangle = a|0\rangle + b|1\rangle$ and see what happens to it under the action of this type of decoherence (phase decoherence). Defining the operator $R(\phi) = \exp[i\phi(I - \sigma_z)/2]$, which has the action $R(\phi)|0\rangle = |0\rangle$ and $R(\phi)|1\rangle = e^{i\phi}|1\rangle$, we have that after the effects of the decoherence, the density matrix of the qubit is given by

$$\rho = \int_0^{2\pi} d\phi \, p(\phi) R(\phi) |\psi\rangle \langle\psi| R^\dagger(\phi)$$
$$= |a|^2 |0\rangle\langle 0| + |b|^2 |1\rangle\langle 1| + a^* b z |1\rangle\langle 0| + a b^* z^* |0\rangle\langle 0|, \qquad (9.56)$$

where

$$z = \int_0^{2\pi} d\phi \, p(\phi) e^{i\phi}. \qquad (9.57)$$

Since $|z| \leq 1$, and is usually less than one, phase decoherence causes the magnitude of the off-diagonal elements of the initial density matrix to decrease. An extreme case is when the phase is uniformly distributed ($p(\phi) = 1/2\pi$), in which case $z = 0$, and the off-diagonal elements will vanish. This would lead to a complete destruction of the phase information in the initial state $|\psi\rangle$.

Now let us consider two qubits, and we shall assume that they are subject to the same random-phase errors. That means that after the phase decoherence has taken place, the two-qubit state $|\Psi\rangle$ will become

$$\rho = \int_0^{2\pi} d\phi \, p(\phi) R(\phi) \otimes R(\phi) |\Psi\rangle \langle\Psi| R^\dagger(\phi) \otimes R^\dagger(\phi). \qquad (9.58)$$

Note that under the action of this kind of decoherence, the states $|0\rangle|1\rangle \to e^{i\phi}|0\rangle|1\rangle$ and $|1\rangle|0\rangle \to e^{i\phi}|1\rangle|0\rangle$ are affected in the same way. Furthermore, we see that any superposition of them

$$R(\phi) \otimes R(\phi)(a|0\rangle|1\rangle + b|1\rangle|0\rangle) = e^{i\phi}(a|0\rangle|1\rangle + b|1\rangle|0\rangle), \qquad (9.59)$$

is just multiplied by an overall phase. When a state of this type is inserted into Eq. (9.58), the overall phase simply cancels out and the state is unchanged. Therefore, states of the form $(a|0\rangle|1\rangle + b|1\rangle|0\rangle)$ are not affected by phase decoherence

We can take advantage of this fact and protect the state of a single qubit from phase decoherence by encoding it into two qubits in the subspace spanned by $|0\rangle|1\rangle$ and $|1\rangle|0\rangle$. In particular we can encode the single-qubit state $|0\rangle$ as $|0\rangle|1\rangle$ and the single-qubit state $|1\rangle$ as $|1\rangle|0\rangle$. As long as the phase decoherence affects both qubits in the same way, this encoding will ensure that any state of a single qubit will be free from the effects of phase decoherence.

9.4   Problems

1. Show that the three-qubit bit-flip code, $|0\rangle \to |0\rangle^{\otimes 3}$ and $|1\rangle \to |1\rangle^{\otimes 3}$, satisfies the quantum error correction condition for the error sets $\{I, X_1, X_2, X_3\}$ and $\{I, Y_1, Y_2, Y_3\}$, but not for the combined set $\{I, X_1, X_2, X_3, Y_1, Y_2, Y_3\}$.

2. For a nondegenerate $[n, k]$ quantum code, each error maps the code space into a different subspace, and all of those subspaces are orthogonal. Suppose we want to be able to correct up to t single-qubit errors, X, Y, or Z.
   (a) Show that

$$\sum_{j=0}^{t} \binom{n}{j} 3^j \leq 2^{n-k}.$$

   This is the quantum Hamming bound.
   (b) Find the smallest value of n allowed by this bound for $k = 1$ and $t = 1, 2$.

3. Find the decoherence-free subspaces for three qubits all undergoing the same random-phase noise errors.

4. One type of code we did not discuss is one that protects against erasure errors. A one-qubit erasure error is equivalent to losing one qubit. We are going to look at the case of qutrits. Consider the following encoding for a qutrit:

$$|0\rangle \to \frac{1}{\sqrt{3}}(|000\rangle + |111\rangle + |222\rangle)$$

$$|1\rangle \to \frac{1}{\sqrt{3}}(|012\rangle + |120\rangle + |201\rangle)$$

$$|2\rangle \to \frac{1}{\sqrt{3}}(|021\rangle + 102\rangle + |210\rangle).$$

   We now use this encoding to encode a general one-qutrit state $|\psi\rangle = \alpha|0\rangle + \beta|1\rangle + \gamma|2\rangle$ into a three qutrit state.

   (a) Show that if we lose two of the qutrits, there is no information about the state $|\psi\rangle$ remaining.
   (b) Show that if we only lose one of the qutrits, we can perfectly recover the state $|\psi\rangle$.

References

1. J. Preskill, *Lecture notes for Physics 219: Quantum Computation.* http://www.theory.caltech.edu/people/preskill/ph229/
2. D.A. Lidar, K.B. Whaley, Decoherence-free subspaces. In *Irreversible Quantum Dynamics*, ed. by F. Benatti, R. Floreanini. Lecture Notes in Physics, vol. 622 (Springer, Berlin, 2003), p. 83 and quant-ph/0301032

Index

J.A. Bergou and M. Hillery, *Introduction to the Theory of Quantum Information
Processing*, Graduate Texts in Physics, DOI 10.1007/978-1-4614-7092-2,
© Springer Science+Business Media New York 2013

Printed in the United States
By Bookmasters